Critical Control Systems

THEORY, DESIGN AND APPLICATIONS

ENGINEERING SYSTEMS MODELLING AND CONTROL SERIES

Series Editor: **Professor David H. Owens**
School of Engineering, University of Exeter, England

1. Simulation and Control of Electrical Power Stations
J. B. Knowles

2. Critical Control Systems
THEORY, DESIGN AND APPLICATIONS
J. F. Whidborne *and* **G. P. Liu**

Critical Control Systems

THEORY, DESIGN AND APPLICATIONS

J. F. Whidborne
University of Leicester, UK

and

G. P. Liu
University of York, UK

RESEARCH STUDIES PRESS LTD.
Taunton, Somerset, England

JOHN WILEY & SONS INC.
New York · Chichester · Toronto · Brisbane · Singapore

RESEARCH STUDIES PRESS LTD.
24 Belvedere Road, Taunton, Somerset, England TA11HD

Marketing and Distribution:

Australia and New Zealand:
Jacaranda Wiley Ltd.
GPO Box 859, Brisbane, Queensland 4001, Australia

Canada:
JOHN WILEY & SONS CANADA LIMITED
22 Worcester Road, Rexdale, Ontario, Canada

Europe, Africa, Middle East and Japan:
JOHN WILEY & SONS LIMITED
Baffins Lane, Chichester, West Sussex, England

North and South America:
JOHN WILEY & SONS INC.
605 Third Avenue, New York, NY 10158, USA

South East Asia:
JOHN WILEY & SONS (SEA) PTE LTD.
37 Jalan Pemimpin 05-04
Block B Union Industrial Building, Singapore 2057

Library of Congress Cataloging in Publication Data

Whidborne, J. F. (James Ferris), 1960–
Critical control systems: theory, design, and applications / J. F. Whidborne, G. P. Liu
p. cm.—(Engineering systems modelling and control series; 2)
Includes bibliographical references and index.
ISBN 0-86380-153-6 (Research Studies Press Ltd.).—ISBN 0-471-94183-2 (Wiley)
1. Automatic control. 2. Process control. I. Liu, G. P. (Guo Ping), 1962– . II. Title. III. Series.
TJ213.W437 1993
629.8—dc20 93-11314
CIP

British Library Cataloguing in Publication Data

A catalogue record for this book
is available from the British Library.

ISBN 0 86380 153 6 (Research Studies Press Ltd.)
ISBN 0 471 94183 2 (John Wiley & Sons Inc.)

Printed in Great Britain by Galliard (Printers) Ltd., Great Yarmouth

Dedication

To my parents
(J.F. Whidborne)

To my parents, wife and daughter
(G.P. Liu)

Foreword

Modelling and control systems design for industrial processes is a multidisciplinary activity requiring the combination of plant knowledge and aspects of, for example, data analysis, computer engineering, instrumentation and computational control algorithms. It is, in many ways, one of the most quantitative and mathematical areas of engineering. The use of linear systems methods in both the time and frequency domain is now well-known and is commonly introduced to undergraduates during their degree courses. The methods can be extremely successful in the development of preliminary designs, and certainly form an effective mechanism for the development of control strategies. Having recognised these facts, it is also increasingly realised that considerations of closed-loop systems robustness, the effect of nonlinearities and the inclusion of constraints in the design process are vital if control technology is to match the increasing performance demands made on many modern systems. The importance of these topics also grows if design and operation are to proceed in the presence of uncertainty in the details of plant behaviour. These problems raise many issues of both a general and detailed form and are almost necessarily approached from a model-based mathematical and hence computational point of view.

The range of techniques used is now very wide but several can be found in the following text, which brings together a number of strands of related research to provide a comprehensive review, a detailed analysis and development of a number of specific design issues and approaches aided by an illustration of the application of the methods and concepts.

Professor David H. Owens,
Exeter, May 1993.

Preface

The advent of interactive computing facilities in the early 1970's led to a reassessment of design methods for control systems. Zakian realised that the growing numerical power of computers coupled with the interactive facilities could be harnessed to great effect, and suggested that many feedback control design problems should be posed as the satisfaction of a set of inequalities, rather than the minimisation of some objective function with inequalities acting as side-constraints, and he proposed the method of inequalities (MI) [ZAN73].

The method of inequalities is an interactive multi-objective method for computer-aided design, where the design problem is expressed as a set of inequalities, known as the design criteria. The inequalities consist of a set of performance indices with a corresponding set of performance specifications (bounds); the aim is to satisfy the inequalities so that the performance indices do not exceed their corresponding specifications (bounds). The inequalities can generally be solved by numerical methods. This means that there are few restrictions on the definitions of the performance indices used in the inequalities; and this in turn means that the method is very flexible. Analytical solutions do exist for solving a few particular inequality problems (e.g. [DGKF89]), and some are developed in this book. In addition, analytical optimal solutions can give great insight into a control system.

The original implementation of the method of inequalities for control system design used traditional performance indices based on the system step response, such as rise time, settling time and steady state error, and many successful designs were carried out using the method with these indices (e.g. [Coe79, Tai79, CD82]). Frequency domain performance indices, such as gain and phase margin, bandwidth and closed-loop peak magnitude have also been used [BCS79]. The usefulness of the MI has also been shown by comparative studies [Tai78, Dah81].

However, it soon became clear that the development of control system design was no longer hampered by the lack of efficient mathematical programming methods, but by the inadequacy of design criteria for expressing the real aims of control. The work of Zakian [Zak78, Zak79b, Zak83, Zak84] led to a design framework, called Zakian's framework , for the design of control systems [Zak86a, Zak86b, Zak87a, Zak87b, Zak89, Zak91]. Based on an explicit definition of control, Zakian's framework provides control system designers with a set of appropriate and far-reaching design criteria.

Traditional models of systems are confined to a description of the dynamical system; that is, the relationship between the system inputs and outputs. Such a description of the system is limited, in that it does not describe the external environment which acts on the system in the form of inputs and disturbances. Thus a more complete description of the system includes not only a model of the dynamical system in terms of the input-output rule, but a model of the environment in terms of characterisations of the inputs. In this book, this complete description of the system is called the 'whole system' [Zak92].

Much of control theory is based on the restricted description of a system. Consequently, the concepts of performance (for example, the integral of absolute error), of controllability (see, for example, [Ros70]) and of stability (see, for example, [Wil70]) do not involve an explicit description of the input .

In contrast, Zakian's framework is based on the larger description. By providing a means of accommodating a model of the system's environment, and facilities for an explicit statement of the control objectives, it overcomes two important disadvantages of conventional control approaches. One is that most of the criteria in conventional control approaches specify performance measures with respect to certain test-input functions, e.g. a step, a sinusoid, white noise, which are often not a realistic representation of the environment of the system. The other is that the design objectives are often not stated or formulated explicitly. Since Zakian's framework involves an explicit description of the environment in terms of an input space, the ensuing concepts of performance, controllability and stability are related to the input space [Zak86a]. Further, by laying down a clear relation between the input space, the input-output mapping and the output performance, Zakian's framework provides a sound theoretical framework for assessing the appropriateness and relative merits of any design criterion [Zak86a, Zak86b, Zak87a].

Zakian's framework has been applied to the design of many practical systems, e.g. the brake control of heavy duty trucks [Bad87b], a maglev vehicle suspension control system [Whi93], a hydraulic turbine control system [LWT90], the control of motor vehicles [Rut92]. In addition, it has been extended to multivariable systems [Zak87a, Zak89]. Based on Zakian's framework, some comprehensive, integrated and interactive computer-aided control system design packages have been developed [Bad86, Bad87a, Whi89]. This software is available from the authors.

Out of much of this work came the realisation that there exists an important class of control systems, namely 'critical systems', and Zakian's framework is ideal for their design [Zak89, Zak91]. A critical system is one where it is necessary to keep the error or other outputs within strictly defined bounds for effective operation of the system. This monograph is based on research work by the authors at the Control Systems Centre, UMIST, and describes the theory and design of continuous time and discrete time critical systems. The theory is backed by applications and case-studies. The work is aimed at research workers as well as advanced engineers in industry who may be frustrated by the scarcity of theoretical work into this very important class of control systems.

Monograph Outline

In Chapter 1, critical control systems are defined, and some examples of such systems are briefly described. It contains a general discussion on the control problem, on performance indices and on the importance of choosing the most appropriate performance index. Methods of characterising system inputs are discussed. It summarises the various approaches to control systems design, including the L^1/l^1, LQG, and H^∞-optimisation approaches. The method of inequalities (MI) is introduced. Various performance indices are considered, and an appropriate performance index for critical systems is introduced. The use of the MI with appropriate measures of performance is seen to be suitable for the design of critical systems.

Chapter 2 contains a collection of known results and concepts from linear system theory, Hardy space $\mathbb{H}^p$, Lebesgue space l^p and polynomial theory, all of which are used in this monograph. Some useful results and concepts from functional analysis are also included. The MI is described in detail and a summary of the design procedure is given.

A number of numerical algorithms which are useful for design with the MI are described in Chapter 3. These are the moving boundaries process (MBP), which is used to solve a set of inequalities, the I_{MN} approximants method of calculating time responses, and the repeated Routh test, which is used to test for stability of the system.

In Chapter 4 several kinds of input space which model the environment of discrete control systems are introduced, along with an appropriate performance function and performance criterion for discrete critical systems. The input spaces are classified into two main types: input spaces $\mathbb{L}(m, \delta)$ and $\mathbb{L}(N, m_0, \delta_0)$ which are based on the magnitude of inputs, and input spaces $\mathbb{D}(m, \delta)$ and $\mathbb{D}(N, m_0, \delta_0)$ which are based on the increment of inputs. The performance function for discrete systems is related to input spaces. The performance criterion is expressed as an inequality representing the design requirement of the performance function.

Chapter 5 is devoted to the design of discrete systems with external inputs in the spaces $\mathbb{L}(m, \delta)$ and/or $\mathbb{L}(N, m_0, \delta_0)$, including persistent and/or transient inputs, to satisfy a multi-performance criterion (MPC), which is a set of performance criteria that represent different requirements for different sets of external inputs. A controller satisfying the MPC is called an admissible controller. The construction of optimal solution of the admissible controllers for minimum phase or non-minimum phase systems is given. Some relations between an input space, an admissible controller and output performance are established. It is shown that the l^1-optimisation problem for discrete time is converted into a simple function minimisation problem. The numerical solution of the multi-performance problem is carried out by the method of inequalities.

In Chapter 6 controllers for discrete systems subjected to external inputs in the spaces $\mathbb{D}(m,\delta)$ or $\mathbb{D}(N,m_0,\delta_0)$, called sup controllers, are presented. The sup controllers minimise the supremum of the absolute value of the generalised output for all time and all external inputs in the space $\mathbb{D}(m,\delta)$ or $\mathbb{D}(N,m_0,\delta_0)$. The analysis, which is restricted to minimum phase discrete systems, yields the sup regulator for systems with disturbances but without reference inputs, finds some relations between an input space, a sup controller and output performance, establishes the uniqueness of the sup controllers and considers the stability of the closed-loop systems. The sup controllers for non-minimum phase discrete systems are also studied. It is shown that they can be designed by numerical methods, e.g. minimax optimisation methods. The results of simulations demonstrate the operation of the sup controllers.

Chapter 7 deals with the problem of controlling critical systems with a bound on the increment of their disturbances and with constant but unknown parameters. A modified least squares (MLS) algorithm is developed to estimate the parameters for the above systems. A self-tuning sup regulator algorithm based on the MLS algorithm and the sup regulator, which is described in Chapter 5, is presented and analysed. Under the assumption that the system to be controlled is stably invertible, the global convergence analysis shows that the closed-loop system is stable and the properties of the self-tuning sup regulator asymptotically converge to those of the sup regulator. These results are illustrated by simulation examples.

The design of continuous time critical control systems is described in Chapter 8. Various input spaces are discussed. An appropriate performance index for continuous time critical systems is defined. Resulting concepts of stability and controllability are discussed. Relationships between the input spaces and output performance are derived. These relationships are combined with the MI to provide an integrated framework for the design of continuous time critical systems. Details are given of how the conflicting requirements of plant disturbance rejection and measurement noise rejection are achieved using the design framework.

Chapter 9 describes in detail the design of two actual critical systems. To prevent an aircraft straying from its designated flight-path, an altitude hold autopilot is required to maintain the aircraft to within 50 feet of the desired altitude. The design of the control system for such an autopilot is described. For the effective operation of a magnetically levitated (maglev) vehicle using electromagnetic suspension, it is necessary that the airgap between the guideway and the levitating magnets is maintained. The maximum control voltage must also be restricted, and, to ensure passenger comfort, it is also desirable to restrict the maximum acceleration experienced by the passengers. The design of the suspension system for a high-speed maglev vehicle which ensures that the airgap, the control voltage and the passenger acceleration are maintained within the prescribed bounds is described.

Chapter 10 extends the framework to the design of sampled-data systems in continuous time. Determination of performance between sampling times is considered. A critical system, the position control of a satellite antenna, is designed in continuous time for implementation with a digital controller.

Chapter 11 is concerned with frequency domain design approaches for continuous time critical systems. Two input spaces are considered. Some important and useful relations are established between an input space, a controller and output performance. They show that the problem of designing continuous time critical control systems can be converted into a standard H^∞ synthesis problem, which is solved by H^∞-optimisation methods. Some simulation examples are given to demonstrate the frequency domain approaches.

Finally, a brief note is in order concerning the authors' responsibilities. Chapters 2, 4, 5, 6, 7 and 11 were written mainly by G.P. Liu, whilst Chapters 1, 3, 8, 9 and 10 were written mainly by J.F. Whidborne.

Acknowledgements

We would like to thank Dr V. Zakian who inspired this work, and for his constant advice and guidance. Many thanks also to our colleagues, N.K. Rutland, S. Arunsawatwong and P.G. Lane for their invaluable help and for our many interesting discussions. Thank you also to A. Browne. We are indebted to S. Simpson for her help with the computer systems and to all the other staff of the UMIST Control Systems Centre for their assistance.

J.F. Whidborne,
Department of Engineering,
University of Leicester,
Leicester LE1 7RH, UK.

G. P. Liu,
Department of Electronics,
University of York,
York YO1 5DD, UK.

Contents

List of Figures

List of Tables

Notation

Symbols

The symbols listed here are used unless otherwise stated. Variables which are functions of time are normally represented by lower case letters, and the Laplace and z-transforms of the variables are normally represented by the corresponding capital letters e.g. $X(s) = \mathcal{L}[x(t)]$.

$a := b$	a is defined as b
$\bar{A}_p$	see Section 2.6
C	controller of discrete systems
C_O	the sup controller
C_R	the sup regulator
$\mathbb{D}(m, \delta)$	input space in the discrete time domain — see Definition 4.3
$\mathbb{D}(N, m_0, \delta_0)$	input space in the discrete time domain — see Definition 4.4
δ	Dirac delta function
e	error output
$\mathbb{E}$	external input space
$\\| f \\|_n$	n-norm of the sequence $\{f_i\}, i \in \mathbb{N}^+$
$\mathcal{F}(D)$	input space in the continuous time domain — see Definition 8.1
$\mathcal{F}(m, D)$	input space in the continuous time domain — see Definition 8.2
$\mathcal{F}(N, m_0, D_0)$	input space in the continuous time domain — see Definition 8.3
$\mathcal{F}(M)$	input space in the continuous time domain — see Definition 8.4
$\mathcal{F}(I)$	input space in the continuous time domain — see Definition 8.5
$\mathcal{F}(A)$	input space in the continuous time domain — see Definition 9.1
$\mathcal{F}(F, D)$	input space in the frequency domain — see Definition 8.6
$\mathcal{F}(D_1, D_2)$	input space in the frequency domain — see Definition 8.7
G	plant of continuous systems
h	unit step function
$\mathbb{H}^p$	Hardy space — see Section 2.6
$\mathbb{R}\,\mathbb{H}^p$	see Definition 2.5
$\mathbb{H}^p(\delta)$	see Definition 2.6
$\mathbb{R}\,\mathbb{H}^p(\delta)$	see Definition 2.7
$\\| H \\|_n$	n-norm of the function $H(s)$
$\\| h \\|_n$	n-norm of the function $h(t), t \in \mathbb{R}$
$\inf\{\cdot\}$	infimum

K	controller of continuous systems
k	unit of discrete time
l^p	see Section 2.6
$\mathcal{L}[\cdot]$	Laplace transform operation
$\mathcal{L}^{-1}[\cdot]$	inverse Laplace transform operation
$\mathbb{L}(m,\delta)$	input space in the discrete time domain — see Definition 4.1
$\mathbb{L}(N,m_0,\delta_0)$	input space in the discrete time domain — see Definition 4.2
λ	root of the characteristic polynomial
Λ	set of roots of the characteristic polynomial
$\max\{\cdot\}$	maximum
$\min\{\cdot\}$	minimum
$\mathbb{N}$	integer numbers
$\mathbb{N}^+$	non-negative integer numbers
ω	angular frequency
P	plant of discrete systems
p	design parameter
$\mathcal{P}$	set of design parameters
ϕ	performance index
$\Vert P \Vert_{\bar{A}_n}$	$\bar{A}_n$-norm of the polynomial $P \in \bar{A}_n$
$r(t)$	unit ramp function
$\mathbf{R}$	input-output rule
$\mathbb{R}[q,n]$	see Definition 2.2
$\mathbb{R}(s)$	see Definition 2.4
$\mathbb{R}$	field of real numbers $(-\infty,\infty)$
$\mathbb{R}^+$	field of non-negative real numbers $[0,\infty)$
$\Vert R \Vert_V$	norm of the vector $R \in \mathbb{R}^n$
$\mathrm{Re}\{x\}$	real part of a complex number x
s	Laplace operator
$S_D(P,C)$	standard feedback discrete systems
$S_C(G,K)$	standard feedback continuous systems
sgn (x)	'signum function': sgn(x) is 1, 0, or -1, respectively, according as $x<0,\ x=0,\ x<0$.
$\sup\{\cdot\}$	supremum
T	the sampling interval in a discrete time system
t	time
t_s	time after the sampling instant for sampled-data systems
τ	time variable used in convolutions
u	controller output
v	system output
$v(h)$	system unit step response
$V(s,\delta)$	system input/output transfer function
w	system input
$x^{(n)}(t)$	$\mathrm{d}^n x(t)/\mathrm{d}t$, the nth derivative of x

$x^{(-1)}(t)$	$\int_0^t x(t)\, \mathrm{d}t$, the integral of x
z	z-transform operator
$\mathcal{Z}[\cdot]$	z-transform operation
$\mathcal{Z}^{-1}[\cdot]$	inverse z-transform operation
$\lvert \cdot \rvert$	modulus

Acronyms

The following abbreviations are used in this book.

ARMA	autoregressive moving-average model
ARMAX	autoregressive moving-average model with auxiliary input
BIBO	bounded-input bounded-output
CACSD	computer-aided control system design
CAD	computer-aided design
EDS	electro-dynamic suspension
IAE	integral absolute error
ISE	integral square error
LHP	left half-plane
LQG	linear quadratic Gaussian
LQR	linear quadratic regulator
LTI	linear time-invariant
LTV	linear time-variant
MBP	moving boundaries process
MI	method of inequalities
MIMO	multi-input, multi-output
MLS	modified least squares
MPC	multi-performance criterion
RHP	right half-plane
RISE	root integral square error
SISO	single-input, single-output

Part I

Introduction

Chapter 1

Critical Systems and Control System Design

1.1 Critical Systems

A problem that occurs frequently in control engineering is to control an output v (usually the error) of a system subjected to external inputs w (reference inputs and/or disturbances) so that the absolute value of the output is within a prescribed bound ε, that is,

$$|v(t, w)| \leq \varepsilon, \quad \forall t \in \mathbb{R}. \tag{1.1.1}$$

Any violation of the bound results in unacceptable, and perhaps catastrophic, operation. An example is the control of force on yarns of a loom in textile mills [Kip89]. The force on the yarn is required to remain within specified narrow bounds so that the yarn is not broken, despite disturbances such as quality changes of the yarns at random times. Violation of this requirement results in the loom stopping. Such systems are said to be critical [Zak89].

There is a wide variety of critical systems; a small selection is listed here:

i) Air traffic controllers require aircraft to remain within a specified narrow altitude bound [FPEN86]. This is often achieved by means of an altitude-hold autopilot.
ii) The catalytic converter, which will soon be a legal requirement on motor cars to reduce pollution, only works if the engine's air/fuel ratio is within certain narrow bounds [FPEN86].
iii) To ensure uninterrupted communication, an antenna dish tracking a communications satellite must at all times point accurately at the satellite to within a small tolerance [FP80].
iv) For the effective operation of a magnetically levitated (maglev) vehicle using electro-magnetic suspension, it is necessary that the airgap between the guideway and the levitating magnets is maintained [Whi93].

Examples (i) and (iv) are included as case studies in Chapter 9.

Control systems are subject to external inputs and disturbances. The aim of the designer of a critical system is to ensure that the inputs do not disturb the system beyond the critical level, thereby making the system tolerant to the inputs. Clearly, the effect of disturbances on the system are of prime importance in the design and analysis of critical control systems. It is thus imperative that the chosen design technique considers these inputs, and furthermore, considers them explicitly. In addition, performance measures which are directly related to the problem expressed by (1.1.1) are required. Both these conditions apply to the design methods described in this work, hence these methods are particularly appropriate for the design of critical systems, and the existence of such systems provides the main motivation for the further development of these methods.

The problem of critical systems has been recognised in other works, and various methods have been suggested to cope with it. Statistical criteria for the design of critical systems are described by Andreyev [And69], and are based on maximising the probability that the outputs will remain within the desired bounds when subjected to random inputs, the inputs being described in terms of probability distributions. Another method is the set theoretic approach [GS71, BR71, Sch73] which introduces the concept of a 'target tube', within which the system states or outputs must be maintained whilst subjected to unknown-but-bounded disturbances. A computationally efficient algorithm for designing control systems using the set theoretic approach has been developed [USWG82] and successfully applied to a control system [PHS+88].

The next sections of this chapter consists of a general discussion on the control design problem, on performance indices and on the importance of choosing the most appropriate performance index, and on input characterisation. Then follows an introduction to the various approaches to design, and methods of solving the problem, including the method of inequalities, which is used extensively throughout this work. Finally, a variety of performance indices which explicitly consider the system inputs are introduced, and their utility for critical control system design discussed.

1.2 The Design Problem

The object of feedback control is to improve the performance of a system by measuring various outputs of the system, processing the output measurement signals by means of a controller, and using the resulting signals to drive actuators which affect the behaviour of the system. One of the main tasks of the control engineer is to determine the controller. This is the design problem which is discussed in this chapter.

There are two main aspects to the design of a controller. The first is to determine what is the purpose, or aim, of the control, and hence decide upon appropriate measures of performance which encompass the aim and which can be used to assess the design. The second is to find a controller which achieves an acceptable performance; there are several possible design approaches that may be used to do this.

1.3 Measures of Performance

A means of expressing the design aims is required in order that a control system may be effectively designed. The aim of the design could be expressed as qualitative design objectives; for example, a satisfactory design would require that the closed-loop system be stable, that the control effort be not too large, the effects of noise and disturbances be small, and that the system responds in a particular way to the reference inputs. However, the designer needs to be able to make comparisons between different designs, and needs to be able to know the extent to which a design objective has been achieved, especially as the various design objectives are usually conflicting. It is particularly important to have quantitative measures of performance and stability for computer-aided design, so the computer can distinguish between different designs. Such quantitative measures of performance, called performance indices, need to accurately reflect the design aims and give a measure of the extent to which the desired design aims have been achieved. For example, a control system may be introduced into a greenhouse, the aim of which is to provide the plants with an ideal growing environment without using too much heating fuel. Appropriate performance indices which would reflect this would be the maximum deviation of the temperature from the ideal, the maximum deviation of the humidity from the ideal, and the average fuel consumption rate.

Over the years, a wide variety of performance and stability indices have been proposed, but there is no single all purpose performance index which can serve all design situations. The designer should thus choose a performance index, or set of performance indices, which is most appropriate. The suitability of the performance indices chosen is dependent upon the nature of the system and upon the function it is called on to perform. Specifically, the performance indices chosen need to reflect such things as tracking or regulating accuracy, efficiency, robustness, disturbance rejection, cost etc. Of fundamental importance then, to the design process, is choosing which performance indices should be used in order to measure the performance, so that the performance indices accurately reflect the design purpose.

However, the choice of performance and stability indices is also determined by the ease with which they can be calculated. The traditional stability indices of phase and gain margin were easy to compute, being based on the Bode or Nyquist plots. Graphical methods of design, such as Bode and Nyquist, were very powerful methods before the advent of computers, being comparatively easy to calculate by hand. The phase and gain margin can be directly read off the Bode or Nyquist plot. Further sophistication of the design process came with the advent of the Nichols diagram, which made computation of the closed-loop frequency response easier, and hence the use of bandwidth, closed-loop resonance frequency and peak gain as performance indices.

The development of the root-locus plot meant that the roots of the closed-loop characteristic equation became easy to compute, and led to indices such as the position of the dominant roots of the characteristic equation, damping factor and natural frequency.

The advent of digital computers meant that the closed-loop response could be rapidly calculated and subsequently other performance indices based on the closed-loop time response to a step input were developed. These include rise time, settling time, overshoot and steady state error, as well as the integral absolute error (IAE) and integral square error (ISE).

What all the above performance indices have in common is that they do not explicitly consider the inputs or disturbances to the system, or else they consider merely a step input at the reference, an input which does not necessarily occur in all systems. As a consequence, several performance indices have been developed which explicitly consider the inputs, and are hence more suitable for the majority of control problems. The selection of such performance indices requires even greater consideration, as the nature and characteristics of the disturbances and inputs need to be considered. Such performance indices are considered in more detail in Section 1.6, and the appropriateness of the indices for design of critical systems discussed.

1.4 Characterisation of Inputs

A control system is subjected to inputs imposed upon it by the environment in which it exists; the severity and nature of these inputs is of fundamental importance to the resulting performance of the control system. It is thus important to consider not just the dynamical system, which is modelled by an input-output rule, but to include the environment into the system description. The environment can be modelled by certain characterisations of the input. A system description which includes both an environment model and a dynamical system model is referred to as a 'whole system' [Zak92].

In classical control theory, little attempt is generally made to model the environment. The performance is assessed by considering the response of the system to deterministic test functions e.g. a unit step, a pulse, a sine wave, etc. Whilst such inputs do sometimes occur from the environment, it is rare that all the inputs to a control system are deterministic, it being unusual for at least some of a control system's inputs not to have a degree of uncertainty. There are several ways in which uncertain system inputs can be characterised, and so more accurately model the environment.

Inputs are most commonly characterised as stochastic processes with some probability structure. The most widely used stochastic process is Gaussian white noise with zero mean and covariance not greater than unity, which are considered in the well-known LQG optimisation problem [AM71], H^2-optimisation problem [YJB76] and much of stochastic control theory [Ast70, GS84, CMT87]. It is rare that inputs are actually Gaussian white or even coloured noise, but it is mathematically convenient to characterise them as such, and effective control schemes can be designed by considering the inputs in such a way.

Another method of characterising an input is to identify the frequency range over which the input magnitude is significant. The designer would then try and ensure that the gain of the system is appropriately large or small over that frequency range to respectively either track or reject the input.

In the H^∞-optimisation problem [Fra87, DFT91], it is assumed that the external inputs are a square-integrable signal of unity energy. Such inputs have finite energy and are described as transient as they decay to zero with increasing time.

In the set theoretic approach [GS71, BR71, Sch73, USWG82, PHS+88] and in l^1/L^1-optimisation problems [Vid86, Vid91, DP87c, DP87b], the external inputs are assumed to be persistent, bounded signals without any statistics associated with the uncertainty. Such inputs have been described as 'unknown-but-bounded' [Sch73].

However, a greater flexibility in modelling the inputs is obtained if the input is characterised as belonging to a known function space of inputs. The space is characterised by means of one or more bounds on the inputs belonging to it, such as bounds on the amplitude, the rate-of-change, the energy, or some other property of the input. Additionally, inputs can be characterised which have both persistent components and transient components. The characterisation of inputs as belonging to a known function space of inputs is also mathematically convenient as a number of simple relationships have been determined relating various input spaces to the output supremum performance index introduced in Section 1.6.4.

Another advantage of characterising the inputs as an input space is that less information is required to identify the characterisation than for a stochastic process. Furthermore, the input space characterisation generally contains a far richer range of inputs than a stochastic process. However, in some circumstances it may contain a large number of fictitious inputs, that is inputs that do not occur in practice, and the resulting design will be conservative.

1.5 The Design Approach

Once the performance indices have been chosen, and methods of calculating the indices established, the way in which the design problem is formulated to achieve a satisfactory design must be decided upon. Numerous design methods have been proposed over the years, but there are essentially three approaches to design :

i) Procedure Based Methods

ii) Optimisation

iii) The Method of Inequalities

Many of the design methods proposed have drawn from more than one of these approaches. For example, a procedure based method may use some optimisation in one of the design stages; the above approaches are thus not exclusive. In fact, other studies [BBN90, Ng89, NGK57] have categorised the various design approaches differently. In addition, there are design methods, such as pole-placement, which do not readily fit into any of the above categories.

The following sub-sections describe each of the approaches and discuss the advantages of each. The method of inequalities is the approach used in Zakian's framework.

1.5.1 Procedure Based Methods

Procedure based methods are characterised by not having an integral design objective. Instead, a series of procedures is followed in stages, where at each stage the designer will attempt to satisfy various, sometimes heuristic, design objectives. The designer will also divide the system up into sub-systems and try to satisfy various objectives for each sub-system. These methods are often derived from, and include, the classical open-loop design methods such as Bode and Nyquist, and also include sophisticated multivariable design methods such as the inverse Nyquist array [Ros74] and the characteristic locus method [MK77]. For example, in a typical Bode design, the designer may introduce gain to ensure stability, followed by a phase lead compensator to increase the phase lead, followed by a lag compensator to reduce the bandwidth to reduce the effect of measurement noise.

In a typical procedure based method, the designer will analyse the problem, and then decide what procedure to follow to achieve the first objective. The results of his actions are then calculated, and usually displayed graphically. The designer then decides if that objective has been reached, and whether to reiterate at the current stage or to go on to the next stage. The methodology is generally open ended and relies to a great extent on the designer's experience and intuition.

When these methods are used in CACSD, the computer generally acts as a tool to aid the calculations and the production of the graphs which are used extensively in such methods. However, there have been efforts (e.g. [Muh91]) to include expert systems techniques into packages to help the designer with decision making and further automate the design process.

One of the disadvantages of procedure based methods is that the designer requires a good deal of experience and intuition to make decisions about the design. If an unfamiliar problem arises, the designer may not have any idea as to how to proceed further and improve the performance. It also means that the performance of the final design may be limited by the designer's knowledge and experience. Another drawback is that many of the design objectives are not incorporated directly into the design but are achieved by various "rules-of-thumb".

Another disadvantage is that the system must be broken down into understandable sub-systems and these sub-systems individually designed. This process may result in an overall design which is more complicated than necessary, consisting of a large number of simple controllers, whereas a simpler controller scheme designed using an integral approach may achieve a similar performance. Alternatively, the designer may not achieve the performance that could be obtained by use of an integral approach. In particular, the designer will not know whether a substantial improvement in performance is possible.

One advantage with procedure based methods are that usually the controllers on each sub-system are not very complex, and because the system has been broken down into sub-systems, the designer knows and understands what each control-loop is doing, and hence the whole system retains an understandable form.

1.5.2 Optimisation

The development of digital computer in the 1950's led to the development of optimisation techniques, the objective of which is to minimise a performance index, ϕ. There are essentially two approaches to optimisation, analytic methods and parameter optimisation.

Analytic Methods

In optimal design using analytic methods, the controller which optimises a single performance index, ϕ, is obtained by the analytical solution of the design problem. A great deal of research has gone into finding analytical solutions to design problems, as analytic solutions are generally mathematically neat and elegant. Examples of analytic techniques include linear quadratic regulator (LQR) optimisation [PM82], linear quadratic Gaussian (LQG) optimisation [Mac89] and H^∞-optimisation [Mac89]. The designer sets up the problem in the form for which an analytic solution is known to exist, such as the LQG problem, and then calculates, normally with a computer, the solution. An important characteristic of analytical methods is that once the problem has been formulated, there is no interaction between the designer and the design process until the solution has been calculated, and ϕ minimised.

The main advantage of the approach is that the solution is guaranteed to be optimal in the sense defined. A major drawback is that the controllers designed by analytic methods are often complex, with high-order, and are subsequently expensive or difficult to implement, although the increase in performance obtained with these complex controllers is often marginal. In addition, there are usually restrictions on the controller configuration, for example, a state observer is usually required. To overcome the problem of high-order and complex controllers, the design is often followed by a controller reduction phase to find a less complex controller which gives a performance close to the optimum.

One characteristic of analytical methods is that the performance index is not independent from the approach. Since analytical solutions are currently only available for a few specific performance indices, another important disadvantage of this method is the restriction in expressing the design goal. For example, in the LQG problem, the performance index must be of the form given by (1.6.3) (or the equivalent multivariable formulation), and other performance indices may not be included. Because the particular problem rarely exactly matches the chosen design method, the designer will try to fit the problem to the method. The designer will do this by adding fictitious inputs and weighting matrices so as to get a design which will give a good performance and achieves the actual design goals. In this respect, the method is little different from procedure based methods in that the designer must use his experience and intuition to choose the weighting matrices and fictitious inputs. The designer will then check if the actual design goals have been achieved, usually by means of simulations, and if not, appropriately alter the weighting matrices and fictitious inputs until the designer decides that the design is satisfactory.

If the method does match the design goal and the disturbances can be accurately modelled in the form required by the method, then the resulting designs from analytic methods of optimisation do yield the best possible design.

Parameter Optimisation

In parameter optimisation, the design problem is

$$\min_{p \in \mathcal{P}} \phi(p), \tag{1.5.1}$$

where $p \in \mathcal{P}$ denotes the parameters of the controller to be chosen from a given set $\mathcal{P}$. Here ϕ is a scalar performance index, but can be a weighted sum of a set of performance indices, the weightings chosen to reflect the relative importance of the various indices. Numerous numerical search techniques, for example, the modified simplex method [NM65], have been developed to find solutions to this problem. Many other techniques are described by Fletcher [Fle87].

A major advantage of this method is that the designer can set the structure of the controller and the system configuration. Another advantage of parameter optimisation is that a much greater range of performance indices are permitted than with analytical methods.

A minor disadvantage of this method is that the designer must specify a start point for the search, and has no guarantee that the search will find an optimum point quickly. The more complex the problem, the harder it is to come up with a reasonable start point.

Parameter optimisation problems are generally nonlinear and are often non-convex. The search algorithms therefore often converge to a local minimum, and there is no elegant way of telling whether this minimum is global or local. Even if this minimum is global, the designer does not know to what extent the performance could be improved by using a more complex controller or different controller structure or system configuration.

A serious disadvantage of optimising a scalar performance index ϕ, is that the designer has the problem of deciding what would be suitable weightings to place on the various components of the index. This problem has led to the formulation of the control system design problem as a multi-objective optimisation problem.

Given a set of n performance indices $\phi_1(p), \phi_1(p), \ldots, \phi_n(p)$, the multi-objective optimisation problem is

$$\min_{p \in \mathcal{P}} \{\phi_i(p), \text{ for } i = 1 \ldots n\}. \tag{1.5.2}$$

In most cases, the performance indices are in conflict, so the reduction of one performance index leads to the increase in another. Subsequently, the result of the multi-objective optimisation is known as a Pareto-optimal solution [Par06]. A Pareto-optimal solution has the property that it is not possible to reduce any of the performance indices without increasing at least one of the other performance indices.

A point $p^* \in \mathcal{P}$ is defined as being Pareto-optimal if and only if there exists no other point $p \in \mathcal{P}$ such that

$$\begin{array}{ll} \text{a)} & \phi_i(p) \le \phi_i(p^*) \text{ for all } i = 1, \ldots, n \quad \text{and} \\ \text{b)} & \phi_j(p) < \phi_j(p^*) \text{ for at least one } j. \end{array} \tag{1.5.3}$$

The problem with multi-objective optimisation is that there is rarely a unique Pareto-optimal solution, and in practice, there is a very large set of Pareto-optimal solutions. A number of methods have been proposed to find a finite set of Pareto-optimal solutions [Zad63, Lin76, Lin77]. A strategy has been proposed for choosing the 'best' Pareto-optimal solution [Gie78] and has been applied to a practical control problem [TSGJ79].

1.5.3 The Method of Inequalities

The method of inequalities is a multi-objective formulation of the design problem. It deals directly with a set of performance indices, and with corresponding statements, in the form of inequalities, indicating when performance is satisfactory. In the method of inequalities [ZAN73], all the design specifications are directly expressed in terms of a set of inequalities, known as the design criteria, the satisfaction of which means that the design objectives have been achieved and a satisfactory design has been obtained.

The design problem is expressed as

$$\phi_i(p) \le \varepsilon_i \qquad \text{for} \qquad i = 1 \ldots n, \tag{1.5.4}$$

where the performance indices $\phi_i : p \mapsto \phi_i(p)$ map $\mathbb{R}^N$ into $\mathbb{R}$, $p \in \mathcal{P}$ denotes the real vector $(p_1, p_2, \ldots, p_N)$ chosen from a given set $\mathcal{P}$, and ε_i are real numbers chosen by the designer representing tolerable values of the performance indices. The aim here is the satisfaction of the set of inequalities in order that an acceptable design is reached. The above formulation is sometimes referred to as the feasibility problem [BB91].

The method of inequalities also provides a computational procedure for solving the set of inequalities (1.5.4). It is conceptually different from optimisation in that the aim is to achieve satisfactory performance rather than optimal performance. In practice, it also differs from parameter optimisation in that the procedure is interactive; that is, it requires supervision and intervention from the user. The interactive nature of the method of inequalities leaves much of the control over the design process with the designer. The designer can choose the controller structure, and, if a solution to (1.5.4) is not found easily, make decisions about adjusting the bounds ε_i to take account of conflicts between different performance indices $\phi_i(p)$.

The multi-objective nature of the method of inequalities means that several performance indices can be dealt with without having to sum them into a single performance index and having the subsequent problem of deciding what weighting to give the various indices.

The method of inequalities also has several advantages over parameter optimisation. The principal advantage is that the method of inequalities formulates the problem as a set of distinct performance criteria, whereas parameter optimisation either uses a set of performance indices or combines them into a scalar composite index. Because of the interactive nature and the concept of design satisfaction rather than optimisation, the designer can ensure that the design is not conservative and that a more complex, and hence more expensive, controller than is required is not used.

An algorithm for the solution of the method of inequalities called the moving boundaries process [ZAN73] has been developed. This algorithm is simple and effective and has been used to solve a great many problems. Practical experience indicates that it does have finite convergence for a large class of practical problems, but a characterisation of this class has not been established theoretically. It is described in detail in Section 3.2. Polak, Mayne and coworkers have also provided algorithms [BHM79, MPSV82] for the solution of (1.5.4) which have proven convergence for certain classes of problems; these algorithms are more complex than the moving boundaries process.

There are several other multi-objective methods, such as the 'Goal – Attainment Method' [Gem74, FP86], the 'Vector Optimisation Method' [KS79, KS83] and the method due to Nye and Tits [NT86]. Some such methods are summarised by Maciejowski [Mac89] and Ng [Ng89].

The method of inequalities has also been combined with H^∞-optimisation to obtain a multi-objective formulation of an analytical optimisation problem [WPG92]. This approach effectively uses the MI to design the weighting functions required in H^∞-optimisation design.

The method of inequalities, being a multi-objective formulation, means that a wide variety of performance indices can be incorporated into the design process and so achieve an effective design. The use of design criteria incorporated in expression (1.5.4) is a meaningful way of expressing the problem and naturally expresses the design specifications. The interactive nature of the method of inequalities provides the designer with flexibility in making informed design choices.

1.6 Performance Indices

Very few control systems are isolated from the environment in which they operate. Most are subjected to inputs and noise which disturb the system. The effects of these inputs on the system are important to the effectiveness of the control system, so it is important to consider the system inputs in control system design. Traditional performance indices, such as phase and gain margin, do not explicitly consider the system inputs. In order to check the effectiveness of the control system in following the reference input or rejecting disturbances, simulations are performed on the design using inputs which may occur in practice. If the design is ineffective, the system is redesigned and the process repeated until an effective design is reached. A great advantage can obviously be obtained in the efficiency of the design task if the system inputs are explicitly considered in the chosen performance indices.

This section briefly describes several performance indices which explicitly consider the system inputs. Several of these indices, namely the H^2, H^∞, L^1 and l^1 -norms are norms on the system's transfer function (for a more detailed discussion on these norms, see [DFT91, BB91]). These indices are only considered here for the single-input single-output case, but all can be simply extended for the multivariable case.

1.6.1 The Variance of the Output

A well known and widely used performance index is the variance of the error $E\{e^2(t)\}$ or of other outputs. The variance of the error is also known as the mean squared value, and is defined in the time domain as

$$E\left\{e^2(t)\right\} := \lim_{T\to\infty} \frac{1}{2T}\int_{-T}^{T} e^2(t)\ \mathrm{d}t, \tag{1.6.1}$$

which is equivalent in the frequency domain to

$$E\left\{e^2(t)\right\} = \frac{1}{2\pi}\int_{-\infty}^{\infty} \Phi_{ee}(\omega)\ \mathrm{d}\omega, \tag{1.6.2}$$

where $\Phi_{ee}(\omega)$ is the spectral density of the error and is the Fourier transform of its autocovariance function $r_{ee}(\tau) = E\{e(t)e(t+\tau)\}$. The disturbances to the system are often modelled as white or coloured Gaussian noise.

The root mean square (rms) value of the output is sometimes taken as the performance index, this being simply the square root of the variance.

The performance index (1.6.1) is usually adapted to include the control u and the error into a single performance index which is then used in an analytic optimisation method, the LQG problem. For single-input single-output systems, the performance index used is

$$\phi = E\left\{e^2(t) + \rho u^2(t)\right\}, \tag{1.6.3}$$

where ρ is a weighting. The method can also be extended to multivariable systems. The LQG problem has been known for many years and has been widely used to design control systems. It is described in Maciejowski [Mac89].

When the input is white noise, the H^2-norm of the system transfer function G measures the rms of the output, where

$$\| G \|_2 := \left(\frac{1}{2\pi} \int_{-\infty}^{\infty} |G(j\omega)|^2 \, d\omega \right)^{1/2}, \tag{1.6.4}$$

so the LQG problem is sometimes referred to as the H^2-optimisation problem.

The use of the variance of the outputs as a performance index is inappropriate for use with critical systems. This is because a similar weighting is given to an error just outside the bound as to one just within the bound, even though the consequences of the two errors are very different. Furthermore, using variance as a performance index will weigh an error far outside the bound much more heavily than one just outside the bound, although both are equally undesirable for a critical system.

1.6.2 The H^∞-Norm

Another performance index which explicitly considers the inputs is the H^∞-norm, which is the energy gain of the system. This performance index has recently received a lot of attention as a result of the development of the H^∞-optimisation design method.

The H^∞-norm of a single-input single-output transfer function G is denoted by $\| G \|_\infty$, and is the maximum, over all frequencies, of the largest gain of $G(j\omega)$; that is,

$$\| G \|_\infty = \sup \{|G(j\omega)| : \omega \in \mathbb{R}\}. \tag{1.6.5}$$

The H^∞-norm can also be simply expressed for multi-input multi-output systems.

Suppose that it is only known that the input disturbance $w(t)$ has a finite energy denoted by the norm $\| w \|_2$ where

$$\| w \|_2 := \sqrt{\left\{ \int_{-\infty}^{\infty} w^2(t) \, dt \right\}}. \tag{1.6.6}$$

This norm is also known as th L^2-norm. The greatest increase in energy that can occur between the input and the output $y(t)$ is shown by [Vid85a] to be

$$\sup \left\{ \frac{\| y \|_2}{\| w \|_2} : \| w \|_2 = 1 \right\} = \| G \|_\infty, \tag{1.6.7}$$

provided $\| w \|_2 < \infty$ and $G(s)$ is stable, proper and has no poles on the imaginary axis. Thus the H^∞-norm represents a bound on the maximum energy of the outputs over the class of all possible inputs with unit energy.

Similarly, if the square root of the average power of $w(t)$ is defined as

$$\text{pow}\ (w) := \left(\lim_{T\to\infty} \frac{1}{2T} \int_{-T}^{T} u^2(t)\ \mathrm{d}t \right)^{1/2}, \tag{1.6.8}$$

then [DFT91]

$$\sup \left\{ \frac{\text{pow}\ (y)}{\text{pow}\ (w)} : \text{pow}\ (w) = 1 \right\} = \| G \|_\infty . \tag{1.6.9}$$

A major merit of the H^∞-norm is that it has a simple meaning in the frequency domain as well as in the time domain. Thus, the H^∞-norm is a useful performance index to reduce the effects of inputs and disturbances on error and control signals when information is known about the frequency distribution of the inputs.

These factors, and the fact that analytical methods have been developed for minimising the H^∞-norm, mean that H^∞-optimisation and design have received a great deal of attention over recent years. The H^∞-optimisation problem is described by Maciejowski [Mac89], Francis [Fra87] and by Doyle *et al* [DFT91]. The technique has been successfully used on practical problems [YP90].

1.6.3 The L^1/l^1-Norm

The problem of minimising the L^1-norm of the error impulse response for continuous time (and the equivalent l^1-norm of the error impulse response for discrete time) systems has recently received attention [Vid86] and solutions to the problem found [DP87b, DP87c, DP87a]. These results are important for the design of certain classes of critical systems (see Sections 5.2 and 8.5.4). The L^1-norm of a single-input single-output transfer function G with inverse Laplace transform $g(t, \delta)$ is denoted by $\| g \|_1$, and is defined as

$$\| g \|_1 := \int_0^\infty |g(t, \delta)|\ \mathrm{d}t. \tag{1.6.10}$$

Suppose that it is only known that the input disturbance $w(t)$ has a maximum amplitude denoted by the norm $\| w \|_\infty$ where

$$\| w \|_\infty := \sup \{ |w(t)| : t \in \mathbb{R} \} . \tag{1.6.11}$$

This norm is also known as the L^∞-norm on a signal. The L^1-norm on a linear, time-invariant, causal system is an upper bound on the L^∞-norm/L^∞-norm system gain

$$\sup \left\{ \frac{\| y \|_\infty}{\| w \|_\infty} : \| w \|_\infty = 1 \right\} = \| g \|_1 . \tag{1.6.12}$$

1.6.4 Least Upper Bound on the Absolute Error

A natural measure of performance, ϕ_e, is the maximum absolute error that occurs. The lack of performance of a feedback system is hence defined [Zak86a] as the least upper bound of the absolute error (henceforth known as the error supremum), i.e.

$$\phi_e(p) = \sup\left\{|e(t,w,p)| : t \geq 0, w \in \mathcal{F}\right\}, \tag{1.6.13}$$

where $e(t,w,p)$ is the system error, t is time, w is the system input, which is known to the extent that it belongs to the known function space of inputs $\mathcal{F}$, and p is the design vector.

Roughly speaking, $\phi_e(p)$ is the greatest error as time ranges over the half line $[0,\infty)$ and the input ranges over the space $\mathcal{F}$. This is a very natural measure of performance and it has an obvious physical meaning, that being the maximum error experienced by the system, and is the natural performance index for use in critical system design.

Clearly, from (1.6.12), this measure of performance is closely related to the L^1-norm introduced in the previous section (compare with Theorem 8.5). However, using the L^1-norm to assess performance is restrictive, because only a restricted class of inputs and a restricted class of systems can be considered.

Many systems also require other states or outputs to be bounded, such as the control effort u to prevent, for example, actuator saturation. Thus other similar performance indices can be introduced, such as $\phi_u(p)$ where

$$\phi_u(p) = \sup\left\{|u(t,w,p)| : t \geq 0, w \in \mathcal{F}\right\}. \tag{1.6.14}$$

The advantages of this measure of performance for evaluating the effectiveness of critical systems means that it is the performance index used throughout this work.

Chapter 2

Preliminaries

2.1 Introduction

This chapter introduces a number of well-known results in control theory which are particularly applicable to the work in the following chapters of this monograph. The results stated can be considered to be purely preliminary in nature. In addition, the method of inequalities, introduced in Section 1.5.3, is described.

In Section 2.2, the ARMAX model is revised. Section 2.3 considers two standard feedback systems. One is for design in the discrete time domain, the other is a continuous time system for design in the time or frequency domain. In Section 2.4, we state descriptions for control systems. The subject of coprime polynomials is introduced in Section 2.5. Section 2.6 introduces discrete time domain spaces and some norms are defined. The results in this section provide a necessary basis for Chapters 4 - 7. Section 2.7 similarly introduces continuous time domain spaces and norms. Section 2.8 introduces frequency domain spaces of real-rational transfer functions and definitions of $\mathbb{H}^p$ norms are stated. Input-output stability of closed-loop systems is introduced in Section 2.9. Finally, the method of inequalities is described in Section 2.10.

2.2 ARMAX Model

In this monograph, we will assume that all plants to be controlled in the discrete time domain are linear time-invariant (LTI). The input-output characteristics of a general plant are described by the model:

$$B_d y(k) = q^{-d} B_n u(k) + D_n \omega(k), \tag{2.2.1}$$

where the polynomials B_d, B_n and D_n are

$$\begin{aligned} B_d &= 1 + b_{d1}q^{-1} + \ldots + b_{dn}q^{-n}, \\ B_n &= b_{n0} + b_{n1}q^{-1} + \ldots + b_{nm}q^{-m}, \\ D_n &= 1 + d_{n0}q^{-1} + \ldots + d_{nl}q^{-l}, \end{aligned}$$

$y : \mathbb{N} \to \mathbb{R}$, $u : \mathbb{N} \to \mathbb{R}$, $\omega : \mathbb{N} \to \mathbb{R}$ and $d \in \mathbb{N}^+ \backslash 0$ are the output, the input, the disturbance and the time delay of the plant, respectively; for $\lambda \in \mathbb{N}^+$, $q^{-\lambda}$ is a backward shift operator, i.e. $q^{-\lambda}x(k) = x(k-\lambda)$; n, m and l are the orders of the polynomials B_d, B_n and D_n, respectively.

The shorthand notation ARMAX (autoregressive moving-average model with auxiliary input) is used for the model (2.2.1). A similar notation ARMA is used for the model (2.2.1) when $\omega(k) = 0$, that is

$$B_d y(k) = q^{-d} B_n u(k). \tag{2.2.2}$$

The ARMAX provides a convenient means of describing the behaviour of a linear discrete plant. It can simply be taken as the valid description of the input-output properties of the plant under study.

2.3 Standard Feedback Systems

In this monograph we consider two types of LTI standard feedback system: the standard feedback discrete system and the standard feedback continuous system. The majority of feedback control problems can be cast into these forms which can include consideration of the effects of reference input, plant disturbances and measurement noise.

In the discrete time domain, the standard feedback discrete system $S_D(P, C)$ is shown in Figure 2.1. In this figure, r is the reference input, ω is the disturbance, u is the control signal, y is the output, $P : (u, \omega) \mapsto y$ represents the plant and $C : (r, y) \mapsto u$ represents the controller. Let w denote the externally applied input which is either the reference input r or the disturbance ω and let v denote the generalised output which is either the output y or the error $e = r - y$.

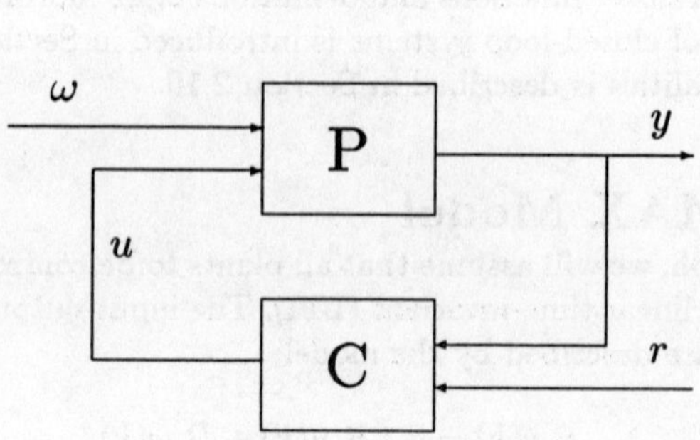

Figure 2.1: The standard feedback system $S_D(P, C)$

The standard feedback continuous system $S_C(G, K)$ is shown in Figure 2.2, where G is a plant, K is a controller, r, ω and ψ are the reference input, the disturbance and the measurement noise, respectively. y is the system output, and u is the control. As for the system $S_D(P,C)$, let w denote the externally applied input which is either the reference input r, the disturbance ω or the measurement noise ψ and let v denote the generalised output which is either the output y or the error $e = r - y$.

Note that if there is no reference input, then the design of the systems $S_D(P,C)$ and $S_C(G,K)$ is a regulation problem, otherwise the design of the systems $S_D(P,C)$ and $S_C(G,K)$ is a control problem. In fact, the regulation problem is a special case of the control problem.

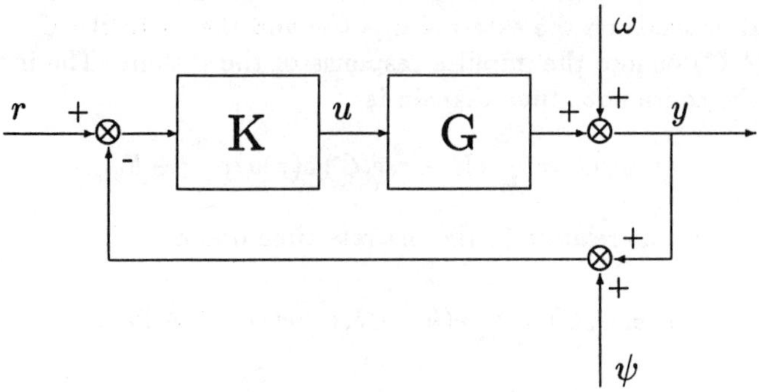

Figure 2.2: The standard feedback system $S_C(G,K)$

2.4 Control System Descriptions

A useful preliminary concept is that of well posedness.

Definition 2.1 *The systems $S_D(P,C)$ and $S_C(G,K)$ are well posed if the transfer functions from the external inputs to the generalised outputs exist and are proper.*

It is assumed from now on that the systems $S_D(P,C)$ and $S_C(G,K)$ are well posed. A control system can be represented by a pair $(\mathbf{R}, \mathbb{E})$, where $\mathbb{E}$ is an external input function space and $\mathbf{R}$ is an input-output rule that maps the external input function space $\mathbb{E}$ into an output function space.

In the time domain $\mathbf{R}$ is expressed by

$$\mathbf{R} : w \mapsto v(w, C), \tag{2.4.1}$$

where w, v and C are the external input, the generalised output and the controller, respectively.

The external input function:

$$w : t \mapsto w(t) \tag{2.4.2}$$

maps $\mathbb{R}$ into $\mathbb{R}$ in the continuous time domain or $\mathbb{N}$ into $\mathbb{R}$ in the discrete time domain.

The generalised output function:

$$v(w, C) : t \mapsto v(t, w, C) \tag{2.4.3}$$

maps $\mathbb{R}$ into $\mathbb{R}$ in the continuous time domain or $\mathbb{N}$ into $\mathbb{R}$ in the discrete time domain, and depends on the external input w and the controller C.

Let $v(t, \delta, C)$ denote the impulse response of the system. The input-output relation in the continuous time domain is

$$v(t, w, C) = \int_0^t v(t - \tau, \delta, C) w(\tau) d\tau, \quad t \in \mathbb{R}^+ . \tag{2.4.4}$$

The input-output relation in the discrete time domain is

$$v(k, w, C) = \sum_{i=0}^{k} v(k - i, \delta, C) w(i), \quad k \in \mathbb{N}^+ . \tag{2.4.5}$$

In the frequency domain, $\mathbf{R}$ is expressed by

$$\mathbf{R} : W \mapsto V(W, K). \tag{2.4.6}$$

where W and V are the Laplace transforms of the external input w and the generalised output v, and K represents the controller.

The input-output relation is defined by

$$V(s, W, K) = V(s, \delta, K) W(s), \tag{2.4.7}$$

where $V(s, W, K)$, $V(s, \delta, K)$ and $W(s)$ are the Laplace transforms of $v(t, W, K)$, $v(t, \delta, K)$ and $w(t)$.

2.5 Coprime Polynomials

In this section we formally define coprime polynomials and outline some important properties associated with coprime polynomials. The results given here have been restricted to the realm of real polynomials and their associated ARMAX models.

Definition 2.2 *Let $\mathbb{R}[q, p]$ denote the set of polynomials in the indeterminate q^{-1} with coefficients in the field $\mathbb{R}$ of real numbers and with the order p in $\mathbb{N}^+$ of non-negative integer numbers.*

For example, the polynomial $A \in \mathbb{R}[q,n]$, that is, $A = a_0 + a_1 q^{-1} + ... + a_n q^{-n}$.

Definition 2.3 *Two polynomials $\alpha \in \mathbb{R}[q,p_1]$ and $\beta \in \mathbb{R}[q,p_2]$ are said to be coprime if their greatest common divisor is 1 (equivalently, they have no common zeros).*

Proposition 2.1 *The polynomials $\alpha \in \mathbb{R}[q,p_1]$ and $\beta \in \mathbb{R}[q,p_2]$ are coprime if and only if there exist polynomials ζ and ξ such that*

$$\alpha\zeta + \beta\xi = 1. \tag{2.5.1}$$

Proposition 2.2 [Ast70] *If the polynomials $B_d \in \mathbb{R}[q,n]$ with $b_{d0} = 1$ and $D_n \in \mathbb{R}[q,l]$ with $d_{n0} = 1$, and $d \in \mathbb{N}^+ \backslash 0$, then there exist two unique polynomials $\alpha \in \mathbb{R}[q,d-1]$ with $\alpha_0 = 1$ and $\beta \in \mathbb{R}[q,n-1]$ such that*

$$B_d\alpha + q^{-d}\beta = D_n. \tag{2.5.2}$$

Proposition 2.3 [CMT87] *If the polynomials $B_d \in \mathbb{R}[q,n]$ with $b_{d0} = 1$ and $D_n \in \mathbb{R}[q,l]$ with $d_{n0} = 1$, and $d \in \mathbb{N}^+ \backslash 0$, then there exist two unique polynomials $\alpha \in \mathbb{R}[q,d-1]$ with $\alpha_0 = 1$ and $\beta \in \mathbb{R}[q,n]$ such that*

$$\Delta B_d\alpha + q^{-d}\beta = D_n, \tag{2.5.3}$$

where $\Delta = 1 - q^{-1}$.

Proposition 2.4 [Ast70] *If the polynomials $B_d \in \mathbb{R}[q,n]$ with $b_{d0} = 1$ and $B_n \in \mathbb{R}[q,m]$ are coprime, $D_n \in \mathbb{R}[q,l]$ with $d_{n0} = 1$ and $d \in \mathbb{N}^+ \backslash 0$, then there exist two unique polynomials $\alpha \in \mathbb{R}[q,d+m-1]$ with $\alpha_0 = 1$ and $\beta \in \mathbb{R}[q,n-1]$ such that*

$$B_d\alpha + q^{-d}B_n\beta = D_n. \tag{2.5.4}$$

The results of Propositions 2.1 - 2.4 will be used in Chapters 4 - 7.

2.6 Discrete Time Domain Spaces

In this section, we introduce the Lebesgue spaces l^p and the various norms that are used in this monograph.

Given a sequence $\{f_i\}$, we define the following norms on it.

$$\| f \|_p := \left[\sum_{i=0}^{\infty} |f_i|^p\right]^{1/p} < \infty, \quad p \in [1,\infty), \tag{2.6.1}$$

$$\| f \|_\infty := \sup\{|f_i| : i \in \mathbb{N}^+\} < \infty. \tag{2.6.2}$$

The sequence space l^p ($p \in [1,\infty]$) is defined to consist of those sequences $\{f_i\}$ such that $\| f \|_p$ is finite.

It is known that a LTI discrete time system with a unit pulse response $\{h_i\}$ is BIBO-stable if and only if $h \in l^1$. Given such a system, we can associate with its polynomial

$$H(q^{-1}) = \sum_{i=0}^{\infty} h_i q^{-i}. \tag{2.6.3}$$

Let $\bar{A}_p$ denote the set of polynomials in $\mathbb{R}[q, \infty]$ of l^p sequences and we define norms on $\bar{A}_p$. Given $H \in \bar{A}_p$, define

$$\| H \|_{\bar{A}_p} := \left[\sum_{i=0}^{\infty} |h_i|^p \right]^{1/p}, \quad p \in [1, \infty), \tag{2.6.4}$$

$$\| H \|_{\bar{A}_\infty} := \sup\{|h_i| : i \in \mathbb{N}^+\}, \tag{2.6.5}$$

$$\| H \|_\infty := \sup\{|H(e^{j\theta})| : \theta \in [0, 2\pi]\}. \tag{2.6.6}$$

The interpretation of these norms is as follows. Given $H \in \bar{A}_p$, one can associate with it as operator, which maps a sequence $\{f_i\}$ into its convolution with $\{h_i\}$, i.e. $H : h * f$, where $*$ denotes the convolution. Then

$$\| H \|_{\bar{A}_p} = \sup\{\| Hf \|_1 \| f \|_m^{-1} : f \in l^m \setminus 0\}, \tag{2.6.7}$$

where $p^{-1} + m^{-1} = 1$.

For example, for $p = 1, 2$, then

$$\| H \|_{\bar{A}_1} = \sup\{\| Hf \|_1 \| f \|_\infty^{-1} : f \in l^\infty \setminus 0\}, \tag{2.6.8}$$

$$\| H \|_{\bar{A}_2} = \sup\{\| Hf \|_1 \| f \|_2^{-1} : f \in l^2 \setminus 0\}. \tag{2.6.9}$$

In addition, for a vector $x \in \mathbb{R}^n$, we define

$$\| x \|_V := \left[\sum_{k=1}^{n} x_k^2 \right]^{1/2}. \tag{2.6.10}$$

Notice that the norm symbol $\| . \|_V$ is specifically used for vectors in $\mathbb{R}^n$.

2.7 Continuous Time Domain Spaces

In this section, we introduce the equivalent continuous time Lebesgue spaces L^p.

Given a function $\{f\}$, we define the following norms on it.

$$\| f \|_p := \left[\int_0^\infty |f(t)|^p \right]^{1/p} < \infty, \quad p \in [1, \infty), \tag{2.7.1}$$

$$\| f \|_\infty := \sup\{|f(t)| : t \in \mathbb{R}^+\} < \infty. \tag{2.7.2}$$

The space of measurable functions L^p ($p \in [1, \infty]$) is defined to consist of those functions $\{f(t)\}$ such that $\| f \|_p$ is finite.

2.8 Frequency Domain Spaces

In this section we will briefly discuss frequency domain spaces of real-rational transfer functions. It should be noted that this discussion is not intended to be comprehensive; for further details, see, e.g. [Vid85a] and [Fra87]. Some preliminary definitions are now given.

The solution of the problem in the frequency domain is facilitated by the use of some concepts and results from the theory of $\mathbb{H}^p$ (or Hardy) space [Dur70].

A Hardy space $\mathbb{H}^p$ ($p \in [1, \infty]$) consists of all complex-valued functions $H(s)$ of a complex variable s which are analytic in the open right half-plane (RHP), $\mathrm{Re}\{s\} > 0$ and satisfy the condition

$$\| H \|_p := \sup\left\{\left[\frac{1}{2\pi}\int_{-\infty}^{+\infty} |H(\sigma + j\omega)|^p d\omega\right]^{1/p} : \sigma > 0\right\} < \infty, \quad p \in [1, \infty), \tag{2.8.1}$$

$$\| H \|_\infty := \sup\{|H(s)| : Re\{s\} > 0\} < \infty. \tag{2.8.2}$$

By the maximum modulus theorem, we can replace the open RHP in (2.8.1) and (2.8.2) by the imaginary axis:

$$\| H \|_p := \left[\frac{1}{2\pi}\int_{-\infty}^{+\infty} |H(j\omega)|^p d\omega\right]^{1/p} < \infty, \quad p \in [1, \infty), \tag{2.8.3}$$

$$\| H \|_\infty := \sup\{|H(j\omega)| : \omega \in \mathbb{R}\} < \infty. \tag{2.8.4}$$

$\| H \|_p$ ($p \in [1, \infty]$) is called p-norm of the function $H \in \mathbb{H}^p$. For each $p \in [1, \infty]$, the space $\mathbb{H}^p$ is a Banach space. Some preliminary definitions and some associated properties are now given.

Definition 2.4 $\mathbb{R}(s)$ *denotes the set of all real-rational functions with real coefficients.*

Definition 2.5 $\mathbb{R}\mathbb{H}^p$ ($p \in [1, \infty]$) *denotes the set of* $\mathbb{H}^p$ *consisting of real-rational functions, i.e.*

$$\mathbb{R}\mathbb{H}^p = \mathbb{R}(s) \cap \mathbb{H}^p. \tag{2.8.5}$$

Definition 2.6 $\mathbb{H}^p(\delta)$ ($p \in [1, \infty]$) *denotes the set of functions* H *satisfying that*

$$H \in \mathbb{H}^p \quad \text{and} \quad \| H \|_p \le \delta, \tag{2.8.6}$$

where δ *is a finite positive number.*

Definition 2.7 $\mathbb{R}\mathbb{H}^p(\delta)$ ($p \in [1, \infty]$) *denotes the set of functions* H *satisfying that*

$$H \in \mathbb{R}\mathbb{H}^p \quad \text{and} \quad \| H \|_p \le \delta, \tag{2.8.7}$$

where δ *is a finite positive number.*

Proposition 2.5 [Fra87] *H belongs to $\mathbb{R}\mathbb{H}^\infty$ if and only if $H \in \mathbb{R}(s)$ and H is proper ($|H(\infty)|$ is finite) and stable (H has no poles in the closed RHP, $Re\{s\} \geq 0$).*

Proposition 2.6 [Fra87] *$H \in \mathbb{R}\mathbb{H}^2$ if and only if $H \in \mathbb{R}(s)$, H is strictly proper ($|H(\infty)| = 0$) and stable.*

In addition, for a set of piecewise-continuous functions $h : \mathbb{R} \to \mathbb{R}$ in the continuous time domain, define

$$\| h \|_n := \left[\int_0^{+\infty} |h(t)|^n dt \right]^{1/n} < \infty, \quad n \in [1, \infty), \tag{2.8.8}$$

$$\| h \|_\infty := \sup\{|h(t)| : t \in \mathbb{R}^+\} < \infty. \tag{2.8.9}$$

$\| h \|_n$ ($n \in [1, \infty]$) is called n-norm of the function h. We have used the same norm symbol for complex-valued functions, real-valued functions and sequence functions. Context determines which is intended.

Note that for the frequency domain design of continuous systems, upper-case characters are usually used to represent the complex-valued functions and lower-case characters are used in this monograph to represent the real-valued functions. In addition, the following relations between the complex-valued functions and the real-valued functions exist: $H = \mathcal{L}[h]$ and $h = \mathcal{L}^{-1}[H]$, where $\mathcal{L}[\cdot]$ is the Laplace transform and $\mathcal{L}^{-1}[\cdot]$ is the inverse Laplace transform.

2.9 Stability Results

In designing a control system, the most important objective is the stability of the system. This is a minimum requirement for any design. It is common in control system analysis to be concerned with input-output stability. This monograph is concerned with input-output stability of closed-loop systems, and a necessary preliminary, therefore, is clearly to define what is meant by input-output stability.

Definition 2.8 *A system is input-output stable if the output response to a bounded input signal is also bounded.*

In other words, the system will be considered input-output stable if a bounded input always produces a bounded output. By bounded, we mean bounded in magnitude.

Theorem 2.1 *A continuous system is stable if and only if all poles of its transfer function are in the left half-plane (LHP).*

Theorem 2.2 *A discrete system is stable if and only if all poles of its transfer function lie outside the closed unit disc of the complex plane.*

Theorems 2.1 and 2.2 are the fundamental theorems for the stability analysis of continuous and discrete control systems in the monograph.

2.10 The Method of Inequalities

The method of inequalities [ZAN73, Zak79b, Zak87b, PM82, Mac89, Ng89], introduced in Section 1.5.3, is central to Zakian's framework for designing critical systems. The design problem is formulated as the set of inequalities

$$\phi_i(p) \leq \varepsilon_i \quad \text{for} \quad i = 1, 2, \ldots, n \tag{2.10.1}$$

where each inequality represents either a performance criterion or a constraint. Here ϕ_i are real functions of p, representing the performance indices and constraints, usually on the controller parameters. $p \in \mathbb{R}^N$ denotes the real vector $(p_1, p_2, \ldots, p_N)$ and is the controller parameters, and ε_i are the largest tolerable values of the respective ϕ_i. The aim here is the satisfaction of the set of inequalities in order that an acceptable design is reached.

Each inequality $\phi_i(p) \leq \varepsilon_i$ of the set of inequalities (2.10.1) defines a set $\mathcal{S}_i$ of points in the n-dimensional space $\mathbb{R}^n$ such that

$$\mathcal{S}_i = \{p : \phi_i(p) \leq \varepsilon_i\}\,. \tag{2.10.2}$$

The boundary of this set is defined by $\phi_i(p) = \varepsilon_i$. A point $p \in \mathbb{R}^N$ is a solution to the set of inequalities (2.10.1) if and only if it lies inside every set $\mathcal{S}_i$, $i = 1, 2, \ldots, n$ and hence inside the set $\mathcal{S}$ which denotes the intersection of all the sets $\mathcal{S}_i$,

$$\mathcal{S} = \bigcap_{i=1}^{n} \mathcal{S}_i. \tag{2.10.3}$$

$\mathcal{S}$ is called the admissible set and any point p in $\mathcal{S}$ is called an admissible point denoted p_s.

The objective is thus to find a point p such that $p \in \mathcal{S}$. Such a point satisfies the set of inequalities (2.10.1) and is said to be a solution. In general, a point p_s is not unique unless the subset $\mathcal{S}$ is a point in the space $\mathbb{R}^N$. In some cases, there is no solution to the problem, i.e. $\mathcal{S}$ is an empty set. It is then necessary to relax the boundaries of some of the inequalities, i.e. increase some of the numbers ε_i, until an admissible point p_s exists.

A computational procedure has been developed to solve the set of inequalities (2.10.1). The procedure is interactive, in that it requires supervision and intervention from the user. It involves choosing the controller structure, the starting point of the search algorithm, and appropriately adjusting the bounds ε_i. A numerical method known as the moving boundaries process (MBP) which solves the inequalities has been developed and is described in Section 3.2.

The general design procedure is summarised below.

i) Define the plant model and parameters, the input space and the bounds ε_i.

ii) Choose the structure of the controller, which also determines the dimension N of vector p, and define the bounds on the elements of p.

iii) Choose the starting point $p \in \mathbb{R}^N$.

iv) Implement a search algorithm, such as the moving boundaries process [ZAN73] to locate a stability point, i.e. a point p such that $\phi_i(p) < \infty$ are satisfied.

v) Starting with a stability point, use the search algorithm to locate a point p which satisfies inequalities (2.10.1).

The interactive nature of the algorithm provides the user with great flexibility. If a solution is not found easily, the designer can change the starting point to avoid computational traps (such as a local minimum of $\phi(p)$), increase the complexity of the controller, or, if appropriate, relax one or more of the bounds ε_i. Alternatively, if a successful design is easily located, and the functionals $\phi_i(p)$ are well within the specified bounds ε_i, the designer may decrease the controller complexity, so giving a cheaper implementation, or may tighten some of the bounds ε_i to obtain a higher quality system.

Chapter 3

Numerical Algorithms

3.1 Introduction

The algorithms described in this chapter can be used to design critical systems using the methods described in this book. The algorithms have been implemented in an interactive computer-aided control system design package known as 'Criteria' [Whi89, Bad87a]. They have also been implemented using the Matlab package [Mat87] which provides a degree of flexibility for CACSD not available within 'Criteria'. [1]

The moving boundaries process is used to solve the method of inequalities. The I_{MN} approximants technique is used to compute time functions and the repeated Routh test provides a procedure for calculating the abscissa of stability, where the abscissa of stability is the maximum real part of the set of closed-loop poles of the system. These algorithms are simple, accurate and efficient and are hence suitable for designing critical control systems.

3.2 Moving Boundaries Process

The method of inequalities, described in Section 2.10, can be solved using a numerical algorithm known as the moving boundaries process (MBP) [ZAN73].

The method of inequalities expresses the design problem as a set of inequalities

$$\phi_i(p) \leq \varepsilon_i \qquad i = 1, \ldots, n, \tag{3.2.1}$$

where ε_i are real numbers, p represents the real vector $(p_1, p_2, \ldots, p_N)$, the elements of which are the controller parameters to be determined, and ϕ_i are real functions of p and represent performance functionals and constraints to be satisfied. The problem is to find a value of p which satisfies all the inequalities (3.2.1).

Each inequality $\phi_i(p) \leq \varepsilon_i$ of the set of inequalities (3.2.1) defines a set S_i of

[1] Both 'Criteria' and the Matlab routines are available from the authors

points in the N-dimensional space $\mathbb{R}^N$ such that

$$\mathcal{S}_i = \{p : \phi_i(p) \leq \varepsilon_i\}. \tag{3.2.2}$$

The boundary of this set is defined by $\phi_i(p) = \varepsilon_i$. A point $p \in \mathbb{R}^N$ is a solution to the set of inequalities (3.2.1) if and only if it lies inside every set $\mathcal{S}_i$, $i = 1, 2, \ldots, n$ and hence inside the set $\mathcal{S}$ which denotes the intersection of all the sets $\mathcal{S}_i$,

$$\mathcal{S} = \bigcap_{i=1}^{n} \mathcal{S}_i. \tag{3.2.3}$$

$\mathcal{S}$ is called the admissible set and any point p in $\mathcal{S}$ is called an admissible point.

The objective is thus to find a point p such that $p \in \mathcal{S}$. Such a point p satisfies the set of inequalities (3.2.1) and is said to be a solution.

The MBP proceeds from an arbitrary initial point to an admissible point, i.e. any point in the set $\mathcal{S}$, in an iterative way. Let p^k denote the value of p at the kth move. $\mathcal{S}_i^k$ is a set formed by the inequality $\phi_i(p) \leq \phi_i(p^k)$ with a boundary $\phi_i(p) = \phi_i(p^k)$. A step is taken from the point p^k to a trial point $\tilde{p}^k$. If for every $i = 1, 2, \ldots, n$, the boundary defined by $\phi_i(p) = \phi_i(\tilde{p}^k)$ is closer, or no further away from, the boundary of $\mathcal{S}_i^k$, then the point $\tilde{p}^k$ is accepted and becomes the new point p^{k+1}. After a sufficient number of successful steps, the boundary of $\mathcal{S}_i^k$ coincides with the boundary of $\mathcal{S}_i$ for every $i = 1, 2, \ldots, n$, and the problem is solved. Thus

$$\mathcal{S}^k = \bigcap_{i=1}^{n} \mathcal{S}_i^k, \tag{3.2.4}$$

$$\mathcal{S}_i^k = \left\{p : \phi_i(p) \leq \varepsilon_i^k\right\}, \qquad i = 1, 2, \ldots, n, \tag{3.2.5}$$

$$\varepsilon_i^k = \begin{cases} \varepsilon_i & \text{if } \phi_i(p^k) \leq \varepsilon_i, \quad i = 1, 2, \ldots, n, \\ \phi_i(p^k) & \text{if } \phi_i(p^k) > \varepsilon_i, \quad i = 1, 2, \ldots, n. \end{cases} \tag{3.2.6}$$

A step is taken from p^k to a trial point $\tilde{p}^k$. This point is a success and we set $p^{k+1} = \tilde{p}^k$ if and only if

$$\phi_i(\tilde{p}^k) \leq \varepsilon_i^k, \qquad i = 1, 2, \ldots, n. \tag{3.2.7}$$

If any of the inequalities (3.2.7) do not hold, another trial point is made from p^k until a success occurs. When all the strict inequalities (3.2.7) hold, the boundaries of the set $\mathcal{S}^{k+1}$ would have moved closer to the boundaries of $\mathcal{S}$. The process is terminated when, after a sufficient number of successful steps, the boundaries of $\mathcal{S}^k$ converged to those of $\mathcal{S}$, i.e. when

$$\varepsilon_i^k = \varepsilon_i, \qquad i = 1, 2, \ldots, n. \tag{3.2.8}$$

3.2.1 Computation

A trial point $\tilde{p}^k$ is generated using a scheme provided by Rosenbrock [Ros60], although other techniques, such as the modified simplex method provided by Nelder and Mead [NM65], have been used [Ng89].

Let V_j^r, $j = 1, 2, \ldots, N$ denote the orthonormal unit vectors, and e_j, $j = 1, 2, \ldots, N$ corresponding real numbers. A trial point is taken as

$$\tilde{p}^k = p^k + e_j V_j^r. \tag{3.2.9}$$

If $\tilde{p}^k$ is a success, e_j is replaced by $3e_j$. If $\tilde{p}^k$ is not a success, e_j is replaced by $-0.5e_j$. In either case j is replaced by $j+1$ in (3.2.9). One iteration of the MBP is defined as N consecutive trials and hence, after the last trial (i.e. $j = N$) of the nth iteration, we let $j = 1$ and $k = k + 1$ for the next iteration. As soon as one success followed by one failure has occurred for every $j = 1, 2, \ldots, N$, vector V_j^r is replaced by V_j^{r+1}, $j = 1, 2, \ldots, N$.

The vectors V_j^{r+1} are computed as follows. Let d_j be equal to the sum of all successful values of e_j during the rth stage, and let

$$\begin{aligned}
a_1 &= d_1 V_1^r + d_2 V_2^r + \cdots + d_N V_N^r, \\
a_2 &= d_2 V_2^r + \cdots + d_N V_N^r, \\
&\vdots \\
a_N &= d_N V_N^r.
\end{aligned} \tag{3.2.10}$$

The Gram-Schmidt procedure is used to orthogonalise the vectors a_i to get

$$\begin{aligned}
b_1 &= a_1 \\
V_1^{r+1} &= b_1 / \| b_1 \|, \\
b_2 &= a_2 - (a_2, V_1^{r+1}) V_1^{r+1}, \\
V_2^{r+1} &= b_2 / \| b_2 \|, \\
&\vdots \\
b_N &= a_N - \sum_{k=1}^{N-1} (a_N, V_k^{r+1}) V_k^{r+1}, \\
V_N^{r+1} &= b_N / \| b_N \|,
\end{aligned} \tag{3.2.11}$$

where (x, y) denotes the inner products of two N-dimensional vectors x and y, and $\|x\|$ denotes the length of the vector x.

Initially at $r = 0$, e_j and V_j^r are chosen arbitrarily. However, as r increases, the rate of convergence of $\mathcal{S}^k$ towards $\mathcal{S}$ tends to improve because V_1^r becomes more orientated along the direction of rapid advance, V_2^r along the direction normal to V_1^r and so on.

3.2.2 Algorithm

i) Data given as ε_i $(i = 1, 2, \ldots, n)$.
Specify maximum number of iterations allowed N_m.

ii) Set $p = p^0$. Compute $\phi_i(p) \quad (i = 1, 2, \ldots, n)$.
If $\phi_i(p) \leq \varepsilon_i \quad (i = 1, 2, \ldots, n)$, the initial point satisfies the inequalities and the problem is solved.

iii) Set $e_j^0 = 0.1|p_j|$ if $|p_j| \geq 0.1 \quad (j = 1, 2, \ldots, N)$.
Set $e_j^0 = 0.01$ if $|p_j| < 0.1 \quad (j = 1, 2, \ldots, N)$.
Set

$$\begin{aligned}
V_1 &= [\ 1\ 0\ 0\ \ldots\ 0\ 0\ 0\ ], \\
V_2 &= [\ 0\ 1\ 0\ \ldots\ 0\ 0\ 0\ ], \\
&\vdots \\
V_{N-1} &= [\ 0\ 0\ 0\ \ldots\ 0\ 1\ 0\ ], \\
V_N &= [\ 0\ 0\ 0\ \ldots\ 0\ 0\ 1\ ].
\end{aligned}$$

Set $\varepsilon'_i = \phi_i(p)$ if $\phi_i(p) > \varepsilon_i \quad (i = 1, 2, \ldots, n)$.
Set $\varepsilon'_i = \varepsilon_i$ if $\phi_i(p) \leq \varepsilon_i \quad (i = 1, 2, \ldots, n)$.
Set $L = 0$, $r = 0$.

iv) Set $e_j = e_j^0$ and $d_j = 0$.

v) Start a new iteration. Set $L = L + 1$ and $j = 1$.

vi) Generate a trial point $\tilde{p} = p + e_j V_j$.
Compute $\phi_i(\tilde{p}) \quad (i = 1, 2, \ldots, n)$.
Test whether $\phi_i(\tilde{p}) \leq \varepsilon'_i \quad (i = 1, 2, \ldots, n)$.
If success go to (vii). If failure go to (viii).

vii) Set $p = \tilde{p}$, $d_j = d_j + e_j$ and $e_j = 3\,e_j$.
Set $\varepsilon'_i = \phi_i(p)$ if $\phi_i(p) > \varepsilon_i \quad (i = 1, 2, \ldots, n)$.
Set $\varepsilon'_i = \varepsilon_i$ if $\phi_i(p) \leq \varepsilon_i \quad (i = 1, 2, \ldots, n)$.
Test whether $\varepsilon'_i = \varepsilon_i \quad (i = 1, 2, \ldots, n)$.
If yes, the problem is solved. Otherwise go to (ix).

viii) Discard $\tilde{p}$ and set $e_j = -0.5\,e_j$.
Did a success and a failure occur for each current V_j ?
If yes, re-initialise vectors V_j $(j = 1, 2, \ldots, N)$ using (3.2.10) and (3.2.11), set $r = r + 1$ and go to (iv).
If no, go to (ix).

ix) If $j = N$, go to (v).
If $j < N$, set $j = j + 1$ and go to (vi).

3.3 I_{MN} Approximants

To calculate the performance of critical systems, it is necessary to efficiently compute system time responses. The I_{MN} approximants technique [Zak75, ZE78, Ino82] is suitable, being based on numerical inversion of Laplace transforms.

3.3.1 Description

Consider the simplified linear differential equation

$$\frac{dx(t)}{dt} = Ax(t) \quad \text{with} \quad x(0) = x_0 \tag{3.3.1}$$
$$y(t) = cx(t). \tag{3.3.2}$$

An I_{MN} approximant of x evaluated at t is defined as

$$I_{MN}(x,t) = \int_0^\infty x(\lambda t)\delta_{MN}(\lambda)\, d\lambda, \tag{3.3.3}$$

where

$$\delta_{MN}(\lambda) = \sum_{i=1}^{N} K_i \exp(-\alpha_i \lambda). \tag{3.3.4}$$

The constants $\alpha_1, \alpha_n, \ldots, \alpha_N$ and $K_1, K_2, \ldots, K_N$ may be defined by the relation

$$e_{MN}^{-z} = \sum_{i=1}^{N} \frac{K_i}{z+\alpha_i}, \tag{3.3.5}$$

where e_{MN}^{-z} is the (M/N) Padé approximant of e^{-z}. The complex numbers K_i and α_i, $i = 1, \ldots, N$ are thus the poles and partial fraction coefficients of e_{MN}^{-z}.

These constants are tabulated in Zakian and Edwards [ZE78] for $N = 1, \ldots$ $\ldots, 10$ and $M = N - 1$. Table 3.1 lists their values for $N = 4$ and $M = 3$. Note that if N is even, the constants occur in complex conjugate pairs.

	real part	imaginary part
α_1	4.787 193 103 128 468	1.567 476 416 895 205
α_2	3.212 806 896 871 536	4.773 087 433 276 642
K_1	-13.301 539 995 971 500	60.071 732 737 047 610
K_2	11.301 539 995 971 490	-12.471 675 850 250 242

Table 3.1: Constants for full-grade I_{MN} approximants for $N = 4$ and $M = 3$

Substituting (3.3.4) into (3.3.3) gives

$$\begin{aligned} I_{MN}(x,t) &= \sum_{i=1}^{N} K_i \int_0^\infty x(\lambda t) \exp(-\alpha_i \lambda)\, d\lambda \\ &= \frac{1}{t}\sum_{i=1}^{N} K_i \int_0^\infty x(\lambda) \exp\left(\frac{-\alpha_i \lambda}{t}\right) d\lambda. \end{aligned} \tag{3.3.6}$$

Now the Laplace transform is defined as

$$X(s) := \mathcal{L}[x(t)] \tag{3.3.7}$$
$$= \int_0^\infty x(\lambda)\exp(-s\lambda)\, d\lambda, \tag{3.3.8}$$

so (3.3.6) is

$$I_{MN}(x,t) = \frac{1}{t}\sum_{i=1}^{N} K_i X\left(\frac{\alpha_i}{t}\right), \quad t > 0. \tag{3.3.9}$$

The $I_{MN}(x,t)$ approximates $x(t)$ very well, especially for small t, that is

$$x(t) \simeq I_{MN}(x,t). \tag{3.3.10}$$

The accuracy of the $I_{MN}(x,t)$ at small values of t can be exploited [Zak75] to provide accuracy at large values of t by effectively shifting the origin of time after each application of the formula, so that the formula sees that time value as a very small value for t and so maintaining the accuracy. The approximate solution to (3.3.1) is thus obtained in a step-by-step fashion as follows. Let t_r, $r = 0,1,2,\ldots$ denote a sequence of points with $t_0 = 0$, that is

$$t_r = h\,r, \quad r = 0,1,2,\ldots \tag{3.3.11}$$

where h is a constant positive step length.

Let

$$X_{t_r} = \int_0^\infty x(t_r + \lambda)\exp(-s\lambda)\,\mathrm{d}\lambda, \tag{3.3.12}$$

so after the change of variable $t = t_r + \lambda$, where λ is a fictitious axis in time, the Laplace transform of (3.3.1) becomes

$$(sI - A)X_{t_r}(s) = x(t_r). \tag{3.3.13}$$

It is clear that the inverse of the Laplace transform $X_{t_r}(s)$ evaluated at h is equal to $x(t_{r+1})$. The I_{MN} approximant of $x(t_{r+1})$ is

$$\frac{1}{h}\sum_{i=1}^{N} K_i X_{t_r}\left(\frac{\alpha_i}{h}\right). \tag{3.3.14}$$

Using (3.3.13) and (3.3.14) and replacing $x(t_r)$ by x_r, the following recursion is obtained

$$\left(\frac{\alpha_i}{h}I - A\right)X_{i,r} = x_r, \quad i = 1,2,\ldots,N, \tag{3.3.15}$$

$$x_{r+1} = \frac{1}{h}\sum_{i=1}^{N} K_i X_{i,r}, \quad r = 0,1,2,\ldots, \tag{3.3.16}$$

where

$$\begin{aligned} t_{r+1} &= t_r + h, && (3.3.17)\\ X_{i,r} &= X_{t_r}(\alpha_i/h), && (3.3.18)\\ x_r &\simeq x(t_r), && (3.3.19) \end{aligned}$$

and x_{r+1} is the I_{MN} approximant of the inverse Laplace transform of $X_r(s)$ at point h. The scheme is capable of providing good accuracy for both large and small values of t.

When N is even, the constants K_i and α_i occur in complex conjugate pairs. Hence it is sufficient to obtain $N/2$ terms for (3.3.9), which becomes

$$I_{MN}(x,t) = \frac{2}{t}\sum_{i=1}^{N} \mathrm{Re}\,\{K_i X(\alpha_i/t)\} \quad \text{with } N \text{ even} \tag{3.3.20}$$

and the recursion (3.3.16) becomes

$$x_{r+1} = \frac{2}{h}\sum_{i=1}^{N/2} \mathrm{Re}\,\{K_i X_{i,r}\} \quad N \text{ is even}\,, \quad r = 0, 1, 2, \ldots. \tag{3.3.21}$$

If the system is represented by a rational transfer function $G(s)$, a minimal realisation technique is used to derive the equivalent state-space representation from which the responses can then be computed. Good results are generally achieved with $N = 4$ and $M = 3$, and these are the values used for the 'Criteria' package.

3.3.2 Computation

The determination of the vector $X_{i,r}$, $i = 1, 2, \ldots, N$ by solving equation (3.3.15) constitutes the major proportion of computational effort in the numerical inversion of Laplace transforms. Any scheme which leads to a reduction in the computational effort will greatly increase the efficiency of the process. One possible scheme [Zak79b] is to take advantage of the regularity and sparsity of the matrix $[sI - A]$, where A is the companion form matrix derived from the state-space realisation of the transfer function representation of the system.

A proper rational transfer function $G(s)$ is given by

$$G(s) = \frac{N(s)}{D(s)}, \tag{3.3.22}$$

where

$$N(s) = B_{n+1}s^n + B_n s^{n-1} + \ldots + B_2 s + B_1, \tag{3.3.23}$$

$$D(s) = A_{n+1}s^n + A_n s^{n-1} + \ldots + A_2 s + A_1. \tag{3.3.24}$$

$G(s)$ can be expressed as

$$G(s) = \frac{N'(s)}{D'(s)} + d, \tag{3.3.25}$$

where

$$N'(s) = c_n s^{n-1} + c_{n-1}s^{n-2} + \ldots + c_2 s + c_1, \tag{3.3.26}$$

$$D'(s) = s^n + a_n s^{n-1} + \ldots + a_2 s + a_1, \tag{3.3.27}$$

$$d = \frac{B_{n+1}}{A_{n+1}}, \tag{3.3.28}$$

(3.3.29)

and

$$a_i = \frac{A_i}{A_{n+1}} \quad \text{for} \quad i = 1, 2, \ldots, n, \tag{3.3.30}$$

$$c_i = \frac{B_i}{A_{n+1}} - d\, a_i \quad \text{for} \quad i = 1, 2, \ldots, n. \tag{3.3.31}$$

A minimal state-space realisation of the transfer function $G(s)$ can be expressed as

$$\frac{\mathrm{d}x(t)}{\mathrm{d}t} = A\, x(t) + b\, f(t) \quad \text{with} \quad x(0) = 0, \tag{3.3.32}$$

$$y(t) = c\, x(t) + d\, f(t), \tag{3.3.33}$$

where $x(t)$ is the state, $f(t)$ is the input and $y(t)$ is the output. The state matrix A of dimension n is the companion form matrix and is given by

$$A = \begin{bmatrix} 0 & \vdots & I & \\ \cdots & \cdots & \cdots & \cdots \\ -a_1 & -a_2 & \cdots & -a_n \end{bmatrix}, \tag{3.3.34}$$

and the input vector is

$$b = \begin{bmatrix} 0 & 0 & \cdots & 0 & 1 \end{bmatrix}^T, \tag{3.3.35}$$

and the output vector is

$$c = \begin{bmatrix} c_1 & c_2 & \cdots & c_{n-1} & c_n \end{bmatrix}. \tag{3.3.36}$$

The matrix $[sI - A]$, is thus

$$[sI - A] = \begin{bmatrix} s & -1 & 0 & 0 & \cdots & 0 \\ 0 & s & -1 & 0 & \cdots & 0 \\ \vdots & \vdots & \vdots & \vdots & \ddots & 0 \\ a_1 & a_2 & \cdots & \cdots & a_{n-1} & s + a_n \end{bmatrix}, \tag{3.3.37}$$

and can be factored into the following sparse factors :

$$[sI - A] = LU, \tag{3.3.38}$$

where

$$L = \begin{bmatrix} 1 & 0 & \cdots & \cdots & \cdots & \cdots & 0 \\ 0 & 1 & 0 & \cdots & \cdots & \cdots & 0 \\ \vdots & \vdots & \vdots & \ddots & \cdots & \cdots & \cdots \\ 0 & 0 & \cdots & \cdots & 0 & 1 & 0 \\ \lambda_1(s) & \lambda_2(s) & \cdots & \cdots & \cdots & \lambda_{n-1}(s) & 1 \end{bmatrix}, \tag{3.3.39}$$

$$U = \begin{bmatrix} s & -1 & \cdots & \cdots & \cdots & \cdots & 0 \\ 0 & s & -1 & \cdots & \cdots & \cdots & 0 \\ \vdots & \vdots & \vdots & \ddots & \cdots & \cdots & \cdots \\ 0 & 0 & \cdots & \cdots & 0 & s & -1 \\ 0 & 0 & \cdots & \cdots & 0 & 0 & \beta(s) \end{bmatrix}, \tag{3.3.40}$$

where

$$\beta(s) = s + a_n + \lambda_{n-1}(s), \tag{3.3.41}$$
$$\lambda_0(s) = 0, \tag{3.3.42}$$
$$\lambda_j(s) = \frac{1}{s}\left[\lambda_{j-1}(s) + a_j\right], \quad j = 1, 2, \ldots, n-1. \tag{3.3.43}$$

It can be shown that

$$L^{-1} = \begin{bmatrix} 1 & 0 & \cdots & \cdots & \cdots & \cdots & 0 \\ 0 & 1 & 0 & \cdots & \cdots & \cdots & 0 \\ \vdots & \vdots & \vdots & \ddots & \cdots & \cdots & \cdots \\ 0 & 0 & \cdots & \cdots & 0 & 1 & 0 \\ -\lambda_1(s) & -\lambda_2(s) & \cdots & \cdots & \cdots & -\lambda_{n-1}(s) & 1 \end{bmatrix}. \tag{3.3.44}$$

After the change of variable $t = t_r + \lambda$ and taking Laplace transforms, equation (3.3.15) leads to

$$(sI - A)X_{t_r}(s) = x(t_r), \tag{3.3.45}$$

which can be rewritten as

$$UX_{t_r}(s) = L^{-1}x(t_r). \tag{3.3.46}$$

The nth component of $X_{t_r}(s)$ is given by

$$\left[s + a_n + \lambda_{n-1}(s)\right] X_{t_r,n}(s) = -\sum_{j=1}^{n-1} \lambda_j(s) x_{r,j} + x_{r,n}, \tag{3.3.47}$$

while the remaining $(n-1)$ terms are given by

$$X_{t_r,k}(s) = \frac{1}{s} X_{t_r,k+1}(s) + \frac{1}{s} x_{r,k}, \quad k = n-1, \ldots, 2, 1. \tag{3.3.48}$$

Replacing s by (α_i/h) in (3.3.42), (3.3.43), (3.3.46) and (3.3.47) yields the complex vectors

$$X_{i,r} = X_{t_r}(\alpha_i/h). \tag{3.3.49}$$

The major advantage of this scheme is that by using the sparsity and regularity of the companion matrix A, the computationally expensive operation of inverting the complex matrices $[(\alpha_i/h)I - A]$, $i = 1, 2, \ldots, N$ in (3.3.15) is avoided.

3.3.3 Algorithm

i) Data given as a rational transfer function $G(s)$ with

(a) numerator coefficients $c_i \quad (i = 1, 2, \ldots, k)$,

(b) denominator coefficients $d_i \quad (i = 1, 2, \ldots, l)$.

Define the total number of steps N_S and the final time.

ii) Read $I_{3/4}$ approximant constants from memory as

(a) real coefficients $K_i \quad (i = 1, 2)$,

(b) imaginary coefficients $\alpha_i \quad (i = 1, 2)$.

iii) Obtain minimal realisation of $G(s)$ as

(a) companion matrix A with elements $a_{ij} \quad (i, j = 1, \ldots, n)$,

(b) input vector b with elements $b_i \quad (i = 1, 2, \ldots, n)$,

(c) output vector c with elements $c_i \quad (i = 1, 2, \ldots, n)$.

iv) Obtain LU factorisation of matrix $[sI - A]$

(a) matrix U with elements $u_{ij} \quad (i, j = 1, 2, \ldots, n)$,

(b) matrix L with elements $l_{ij} \quad (i, j = 1, 2, \ldots, n)$.

v) Apply the recursion of (3.3.15) and (3.3.21) to obtain the values of $x_i \quad (i = 1, 2, \ldots, N_S)$

$$\left[\frac{\alpha_i}{h} I - A\right] X_{i,r} = x_r,$$

$$x_{r+1} = \frac{2}{h} \sum_{i=1}^{N/2} \mathrm{Re}\{K_i X_{i,r}\} \quad N \text{ is even} .$$

vi) Use the output equation 3.3.2 to compute $y_i \quad (i = 1, 2, \ldots, N_S)$.

3.4 Repeated Routh Test

The use of the Routh test was, until recently, little used in the design of control systems compared to methods such as Bode and Nyquist. This was primarily because Bode and Nyquist are graphical techniques and are well suited to hand computation. Since the advent of digital computers, the Routh test has become considerably easier to calculate and hence has become more widely used in the analysis of stability.

A linear time invariant dynamical system has for its characteristic polynomial $D_0(s)$ where

$$D_0(s) = s^n + a_n s^{n-1} + \ldots + a_2 s + a_1, \tag{3.4.1}$$

where the coefficients a_j are known real numbers. The system is stable if and only if its abscissa of stability satisfies the criterion

$$\alpha_0 < 0, \tag{3.4.2}$$

where α_0 is defined as

$$\alpha_0 = \max\{\mathrm{Re}\{z_i\} : i = 1, 2, \ldots, n\}, \tag{3.4.3}$$

where $z_i, \quad i = 1, 2, \ldots, n$ are the zeros of $D_0(s)$.

An efficient and accurate algorithm, known as the repeated Routh test [Zak79a], computes α_0 without computing the zeros z_i. It consists of an iteration in which, at the kth cycle, the Routh test is performed on the shifted polynomial $D_0(s + x_k)$, where x_k is a real number. The infinite sequence $\{x_k\}$ generated by the iteration converges to α_0. The iteration is summarised below.

The Routh test is applied to $D_0(s)$ to determine whether $\alpha_0 \geq 0$ or $\alpha_0 < 0$. If $\alpha_0 \geq 0$ then determine an upper bound u on α_0; alternatively if $\alpha_0 < 0$ determine a lower bound l on α_0. Thus

$$\begin{array}{ccccccc} 0 & \leq & \alpha_0 & \leq & u & \text{if} & \alpha_0 \geq 0, \\ l & \leq & \alpha_0 & < & 0 & \text{if} & \alpha_0 < 0. \end{array} \tag{3.4.4}$$

Now compute the shifted polynomial

$$D_1(s) = D_0(s + x_1), \tag{3.4.5}$$

where either $x_1 = u/2$ or $x_1 = l/2$. Let α_1 denote the abscissa of stability of $D_1(s)$. Apply the Routh test to $D_1(s)$ giving

$$\begin{array}{ccccccc} \frac{u}{2} & \leq & \alpha_0 & \leq & u & \text{if} & \alpha_1 \geq 0, \\ 0 & \leq & \alpha_0 & < & \frac{u}{2} & \text{if} & \alpha_1 < 0, \end{array} \tag{3.4.6}$$

(similarly on l). The procedure is repeated until α_0 is known to a sufficient degree of accuracy.

More explicitly, let $k = 0, 1, 2, \ldots$, and let $\{x_k\}$ be a sequence of real numbers. Also let

$$D_k(s) = D_0(s + x_k), \tag{3.4.7}$$

$$D_k(s) = \prod_{i=1}^{n}(s - z_{i,k}), \tag{3.4.8}$$

$$\begin{aligned} \alpha_k &= \max\{\mathrm{Re}\{z_{i,k}\} : i = 1, 2, \ldots, n\}, & (3.4.9) \\ &= \alpha_0 - x_k, & (3.4.10) \end{aligned}$$

and l and u are the lower and upper bounds on α_0, that is

$$l \leq \alpha_0, \quad \text{if } \alpha_0 < 0, \tag{3.4.11}$$

$$\alpha_0 \leq u, \quad \text{if } \alpha_0 \geq 0. \tag{3.4.12}$$

The sequence $\{x_k\}$ is generated by the iteration

$$x_{k+1} = x_k + h_{k+1}, \quad x_0 = 0, \tag{3.4.13}$$

$$h_{k+1} = \left|\frac{h_k}{2}\right| \quad \text{if } \alpha_k \geq 0, \ h_0 = u, \tag{3.4.14}$$

$$h_{k+1} = -\left|\frac{h_k}{2}\right| \quad \text{if } \alpha_k < 0, \ h_0 = l. \tag{3.4.15}$$

The sequence $\{x_k\}$ converges to the abscissa of stability α_0.

3.4.1 Computation of Shifted Polynomial

The shifted polynomial $D_0(s+x_k)$ is formed from $D_0(s)$ using Horner's algorithm [Hen74]. By synthetic divisions compute, recursively, the numbers $b_1, b_2, \ldots, b_n$ such that

$$\begin{aligned} D_0(s) &= (s - x_k)P_{n-1} + b_1, \\ P_{n-1} &= (s - x_k)P_{n-2} + b_2, \\ P_{n-2} &= (s - x_k)P_{n-3} + b_3, \\ &\vdots \\ P_2 &= (s - x_k)P_1 + b_{n-1}, \\ P_1 &= (s - x_k) + b_n, \end{aligned} \tag{3.4.16}$$

where the P_j are monic polynomials of degree j. By a process of back substitution, P_j can be eliminated from (3.4.16). Let $s_1 = s + x_k$, and the resulting shifted polynomial is given by

$$D_0(s + x_k) = s_1^n + b_n s_1^{n-1} + \ldots + b_2 s_1 + b_1. \tag{3.4.17}$$

3.4.2 Computation of Upper and Lower Bounds

The upper bound u and the lower bound l on α_0 are obtained by the use of Gershgorin's discs theorem [Frö65]. The theorem actually finds bounds on the eigenvalues of any matrix, rather than the roots of a polynomial, so the companion matrix C to the polynomial $D_0(s)$ must be constructed. The companion matrix C is derived from $D_0(s)$ by

$$C = \begin{bmatrix} 0_{n-1} & \vdots & I_{n-1 \times n-1} \\ \cdots & & \cdots\cdots\cdots \\ -a_1 & \vdots & -a_2 \cdots - a_n \end{bmatrix}. \tag{3.4.18}$$

The matrix C has the roots of $D_0(s)$ as its eigenvalues.

Gershgorin's discs theorem states that the eigenvalues of an $n \times n$ matrix (m_{ij}) lie in the union of the discs

$$|s - m_{ii}| \leq \sum_{\substack{j=1 \\ j \neq i}}^{n} |m_{ij}|, \quad i = 1, 2, \ldots, n, \tag{3.4.19}$$

or alternatively in the union of the discs

$$|s - m_{ii}| \le \sum_{\substack{j=1 \\ j\neq i}}^{n} |m_{ji}|, \quad i = 1, 2, \ldots, n. \tag{3.4.20}$$

It can be shown [Zak79a] that u and l are given by

$$u = \min\{c_1, c_2\}, \tag{3.4.21}$$

$$l = \max\{d_1, d_2\}, \tag{3.4.22}$$

where

$$c_1 = \max\left\{1, \left(-a_n + \sum_{i=1}^{n-1} |a_i|\right)\right\}, \tag{3.4.23}$$

$$c_2 = \max\left\{|a_1|, (1 - a_n), (1 + |a_i|) \text{ for } i = 1, 2, \ldots, n-1\right\}, \tag{3.4.24}$$

$$d_1 = \min\left\{-1, \left(-a_n - \sum_{i=1}^{n-1} |a_i|\right)\right\}, \tag{3.4.25}$$

$$d_2 = \min\left\{-|a_1|, (-1 - a_n), (-1 - |a_i|) \text{ for } i = 1, 2, \ldots, n-1\right\} \tag{3.4.26}$$

3.4.3 Algorithm

Let ε be a small positive real number such that when $|x_k| \le \varepsilon, x_k$ is set to zero. Let β_0 be the desired accuracy for α_0.

i) Data given as polynomial $D_0(s)$ with coefficients $a_i \quad (i = 1, 2, \ldots, n-1)$.

ii) Determine u and l by Gershgorin's discs theorem.
Apply Routh test to $D_0(s)$.
If $\alpha_0 \ge 0$, set $h_0 = u$, if $\alpha_0 < 0$, set $h_0 = l$.
Set $k = 0$ and $x_0 = 0$.

iii) If $\alpha_k \ge 0$, set $h_{k+1} = |h_k/2|$.
If $\alpha_k < 0$, set $h_{k+1} = -|h_k/2|$.

iv) Set $x_{k+1} = x_k + h_{k+1}$.
If $x_{k+1} \le \varepsilon$, set $x_{k+1} = 0$.

v) Evaluate the accuracy parameter β_{k+1}:
if $x_{k+1} > \varepsilon$, set $\beta_{k+1} = |h_{k+1}/x_{k+1}|$,
if $x_{k+1} \le \varepsilon$, set $\beta_{k+1} = |h_{k+1}|$.

vi) Check the accuracy:
If $\beta_{k+1} \le \beta_0$, set $\alpha_0 = x_{k+1}$ and stop.
Else go to (vii).

vii) Set $k = k + 1$.
Compute the shifted polynomial $D_k(s) = D_0(s + x_k)$.

viii) Apply the Routh test to $D_k(s)$.
Go to (iii).

Part II

Discrete Time Systems

Chapter 4

Discrete Time Systems: Input Spaces and Performance Criterion

4.1 Introduction

It was seen in Chapter 1 that a full description of a system should include models of both the environment and the dynamical system. In this work, the environment, consisting of the system inputs, is modelled by characterising the inputs as belonging to an input space; whilst the dynamical system is represented by an input-output rule. This control system representation is formally described in Section 2.4.

The choice of input space is crucial to the definition of performance and directly affects the design method of a system. Several kinds of input spaces in the discrete time domain are presented in this chapter. Input spaces $\mathbb{L}(m,\delta)$ and $\mathbb{L}(N,m_0,\delta_0)$, are characterised by having some bound on the magnitude of the inputs belonging to them, and are defined in Section 4.2. Input spaces $\mathbb{D}(m,\delta)$ and $\mathbb{D}(N,m_0,\delta_0)$, are characterised by having some bound on the increment of their inputs [LZ90a, LZ90b, Liu92a], and are defined in Section 4.3.

As seen in Section 1.3, the measure of performance in control system design is very important, particularly for critical systems. A performance function for critical systems is thus defined in Section 4.4, based on the corresponding input spaces. A critical control system is said to be well behaved if it ensures that during all time and all external inputs, the generalised output remains tolerably small. Therefore, an appropriate performance criterion is also given in Section 4.4.

4.2 Input Spaces $\mathbb{L}(m,\delta)$ and $\mathbb{L}(N,m_0,\delta_0)$

Discrete time domain input spaces $\mathbb{L}(m,\delta)$ and $\mathbb{L}(N,m_0,\delta_0)$ are defined in this section. Inputs in these spaces have some bound on their magnitude.

Definition 4.1 *The input space $\mathbb{L}(m, \delta)$ is a set of sequences w such that $w_k = 0$, for $k < 0$, and for $m \in [1, \infty]$ and $\delta \in (0, \infty)$,*

$$\| w \|_m \leq \delta. \tag{4.2.1}$$

For $m = \infty$, an important feature of the space $\mathbb{L}(\infty, \delta)$ is that it contains elements w that change substantially for all time $t \in \mathbb{N}^+$ within the constraint $\| w \|_\infty \leq \delta$. For example, $\delta \sin(t) \cos(t)$ is an element of $\mathbb{L}(\infty, \delta)$. The inputs considered in the l^1-optimisation problem belong to $\mathbb{L}(\infty, \delta)$ [Vid86, DP87b]. Control systems are frequently subjected to persistent external inputs which can be modelled by elements of $\mathbb{L}(\infty, \delta)$. Therefore, the external input space $\mathbb{L}(\infty, \delta)$ is called a bounded persistent input space.

In contrast, if $1 \leq m < \infty$, then the input space $\mathbb{L}(m, \delta)$ contains elements w that do not change substantially for all time in $\mathbb{N}^+$. Usually, $\mathbb{L}(m, \delta)$ with $m \in [1, \infty)$ is known as a transient input space [Zak87a].

In particular, if $m = 2$, the input space $\mathbb{L}(2, \delta)$ comprises inputs w with finite energy, which is considered in the H^∞-optimisation problem [Zam81, Vid85a, Fra87].

Proposition 4.1 *For $m \in [1, \infty)$ and $\varepsilon \in (0, \infty)$, then $\mathbb{L}(m, \delta) \subset \mathbb{L}(m + \varepsilon, \delta)$.*

Proof: For all $w \in \mathbb{L}(m, \delta)$, if $g = w \| w \|_\infty^{-1}$, then $\| w \|_m = \| w \|_\infty \| g \|_m$. It follows that

$$|g_i| \leq 1, \quad \text{for} \quad i \in \mathbb{N}^+, \tag{4.2.2}$$

$$\| g \|_n \geq 1, \quad \text{for} \quad n \in [1, \infty). \tag{4.2.3}$$

Since $\varepsilon \in (0, \infty)$ and $m \in [1, \infty)$,

$$\begin{aligned} \| w \|_{m+\varepsilon} &= \| w \|_\infty \| g \|_{m+\varepsilon} \\ &= \| w \|_\infty \left[\sum_{i=0}^{\infty} |g_i|^{m+\varepsilon} \right]^{1/(m+\varepsilon)} \\ &\leq \| w \|_\infty \left[\sum_{i=0}^{\infty} |g_i|^{m} \right]^{1/(m+\varepsilon)}, \quad \text{using} \quad (4.2.2) \\ &\leq \| w \|_\infty \left[\sum_{i=0}^{\infty} |g_i|^{m} \right]^{1/m}, \quad \text{using} \quad (4.2.3) \\ &= \| w \|_\infty \| g \|_m. \end{aligned} \tag{4.2.4}$$

Therefore, $\| w \|_{m+\varepsilon} \leq \delta$ which shows that for $m \in [1, \infty)$ and $\varepsilon \in (1, \infty)$, $\mathbb{L}(m, \delta) \subseteq \mathbb{L}(m+\varepsilon, \delta)$. Also, if $m \in [1, \infty)$ and $\varepsilon \in (0, \infty)$, $\mathbb{L}(m, \delta) \neq \mathbb{L}(m+\varepsilon, \delta)$. To verify this, choose a particular sequence $w_i^* = 2^{-1/(m+\varepsilon)}\delta$, for $i = 0, 1$ and $w_i^* = 0$, otherwise. Then $w^* \notin \mathbb{L}(m, \delta)$ but $w^* \in \mathbb{L}(m + \varepsilon, \delta)$. □□

Proposition 4.2 *For $m \in [1,\infty)$, then $\mathbb{L}(m,\delta) \subset \mathbb{L}(\infty,\delta)$.*

The result of Proposition 4.2 is obvious from the definition of $\mathbb{L}(m,\delta)$. Propositions 4.1 and 4.2 show that the input spaces $\mathbb{L}(m,\delta)$ for $m \in [1,\infty]$ have a relation:

$$\mathbb{L}(1,\delta) \subset \ldots \subset \mathbb{L}(2,\delta) \subset \ldots \subset \mathbb{L}(\infty,\delta). \tag{4.2.5}$$

In many practical systems, either the persistent input space or the transient input space cannot effectively describe the external inputs to which the systems are subjected, because an input w is generally made up of several parts, e.g. a persistent part and a transient part. Thus, suppose the input w is the sum of the sequences $w^{(j)}$ $(j = 1,2,...,N)$, i.e.

$$w = \sum_{j=1}^{N} w^{(j)}, \tag{4.2.6}$$

where $w^{(j)}$ belongs to a class $\mathbb{L}(m_{0j},\delta_{0j})$ as defined above. Let

$$m_0 = [m_{01}, m_{02}, ..., m_{0N}], \tag{4.2.7}$$
$$\delta_0 = [\delta_{01}, \delta_{02}, ..., \delta_{0N}]. \tag{4.2.8}$$

Definition 4.2 *The complex input space is defined as*

$$\mathbb{L}(N,m_0,\delta_0) = \left\{ \sum_{j=1}^{N} w^{(j)} : w^{(j)} \in \mathbb{L}(m_{0j},\delta_{0j}), \quad \text{for} \quad j = 1,2,...,N \right\}. \tag{4.2.9}$$

Clearly, the complex input space $\mathbb{L}(N,m_0,\delta_0)$ covers at least a persistent input if one of the elements in m_0 is infinity and at least a transient input if one of elements in m_0 is finite.

For example, assume that a set of inputs is expressed by

$$w(t) = 0.3\cos(k)\,\mathrm{sgn}\,(\alpha(k)) + 0.2(k+1)^{-1}\,\mathrm{sgn}\,(\beta(k)), \qquad \text{for} \qquad k \in \mathbb{N}^+, \tag{4.2.10}$$

where $\alpha(k)$ and $\beta(k)$ are random variables with zero mean; as usual, $\mathrm{sgn}\,(\cdot)$ is the signum function. Then w can be formulated as consisting of two parts, $w^{(1)}$ and $w^{(2)}$, i.e.

$$w^{(1)}(k) = 0.3\cos(k)\,\mathrm{sgn}\,(\alpha(k)), \tag{4.2.11}$$
$$w^{(2)}(k) = 0.2(k+1)^{-1}\,\mathrm{sgn}\,(\beta(k)). \tag{4.2.12}$$

It is obvious that $w^{(1)} \in \mathbb{L}(\infty, 0.3)$ and $w^{(2)} \in \mathbb{L}(2, 0.2565)$. Thus, $w \in \mathbb{L}(2, [\infty,2], [0.3, 0.2565])$. On the other hand, w can also be formulated by the space $\mathbb{L}(\infty, 0.5)$. Clearly, the space $\mathbb{L}(2,[\infty,2],[0.3,0.2565])$ contains less fictitious elements than $\mathbb{L}(\infty,\ 0.5)$. Therefore, $\mathbb{L}(2,[\infty,2],[0.3,0.2565])$ describes the set of inputs w more accurately.

4.3 Input Spaces $\mathbb{D}(m, \delta)$ and $\mathbb{D}(N, m_0, \delta_0)$

Discrete time domain input spaces $\mathbb{D}(m, \delta)$ and $\mathbb{D}(N, m_0, \delta_0)$ are defined in this section. They are defined on the basis of the increment of the inputs rather than the magnitude of the inputs.

Definition 4.3 *The input space $\mathbb{D}(m, \delta)$ is a set of sequences w such that*

$$\sup\left\{\sum_{i=k}^{k+m} |\Delta w(i)| : k \in \mathbb{N}\right\} \leq \delta, \tag{4.3.1}$$

$$|w(k)| < \infty, \quad \text{for} \quad k \in \mathbb{N}, \tag{4.3.2}$$

where $\delta \in (0, \infty)$, $m \in \mathbb{N}^+$ and as usual, Δw is the increment of w, i.e.

$$\Delta w(k) = w(k) - w(k-1). \tag{4.3.3}$$

From (4.3.1), clearly, the input space $\mathbb{D}(m, \delta)$ essentially implies a restriction on the rate of change of input. If $m = 0$, $\mathbb{D}(0, \delta)$ includes elements w that are persistent for all time within the constraint

$$\sup\{|\Delta w(k)| : k \in \mathbb{N}\} \leq \delta. \tag{4.3.4}$$

$\mathbb{D}(0, \delta)$ is known as a persistent input space.

In contrast, if $m = \infty$, $\mathbb{D}(\infty, \delta)$ does not contain elements w which are persistent, but there exist some w such that

$$\lim_{k\to\infty} \Delta w(k) = 0. \tag{4.3.5}$$

$\mathbb{D}(\infty, \delta)$ is known as a transient input space.

Therefore, the input space $\mathbb{D}(m, \delta)$ covers persistent inputs and transient inputs. Some properties of the input space $\mathbb{D}(m, \delta)$ are given by the following propositions.

Proposition 4.3 *If $0 < m < \infty$ and $0 < r \leq m$, then*

$$\mathbb{D}\left(m - r, \frac{\delta}{r+1}\right) \subset \mathbb{D}(m, \delta). \tag{4.3.6}$$

Proof: For $0 < r \leq m$, there exists the following inequality:

$$\begin{aligned}\sum_{i=k}^{k+m} |\Delta w(i)| \;\leq\; & \sum_{i=k}^{k+m-r} |\Delta w(i)| + \sum_{i=k}^{k+m-r} |\Delta w(i+1)| \\ & + \ldots + \sum_{i=k}^{k+m-r} |\Delta w(i+r)|.\end{aligned} \tag{4.3.7}$$

For all $w \in \mathbb{D}(m-r, \delta(r+1)^{-1})$, then we have

$$\sum_{i=k}^{k+m-r} |\Delta w(i)| \leq \frac{\delta}{r+1}, \quad \forall k \in \mathbb{N}. \tag{4.3.8}$$

Substituting (4.3.8) into (4.3.7) gives

$$\sum_{i=k}^{t+m} |\Delta w(i)| \leq (r+1)\frac{\delta}{r+1} = \delta, \quad \forall t \in \mathbb{N}. \tag{4.3.9}$$

Hence, all $w \in \mathbb{D}(m-r, \delta(r+1)^{-1})$ belong to $\mathbb{D}(m,\delta)$. So,

$$\mathbb{D}\left(m-r, \frac{\delta}{r+1}\right) \subseteq \mathbb{D}(m,\delta). \tag{4.3.10}$$

Also, if $0 < m < \infty$, then

$$\mathbb{D}\left(m-r, \frac{\delta}{r+1}\right) \neq \mathbb{D}(m,\delta). \tag{4.3.11}$$

In order to verify this, suppose that $w(k) = 0$ for $k \neq k_0$ and $w(k) = \delta$ for $k = k_0$. Then $w \in \mathbb{D}(m,\delta)$ but $w \notin \mathbb{D}(m-r, \delta(r+1)^{-1})$. □□

Proposition 4.4 *If $0 < m < \infty$, then*

$$\mathbb{D}(m,\delta) \subset \mathbb{D}(m-1,\delta). \tag{4.3.12}$$

The result of Proposition 4.4 is clear from the definition of $\mathbb{D}(m,\delta)$. Propositions 4.3 and 4.4 imply that for $r = 1$,

$$\mathbb{D}\left(m-1, \frac{\delta}{2}\right) \subset \mathbb{D}(m,\delta) \subset \mathbb{D}(m-1,\delta). \tag{4.3.13}$$

Thus, they show how much larger $\mathbb{D}(m-1,\delta)$ is compared to $\mathbb{D}(m,\delta)$.

We now define the following complex input space which is similar to the complex input space $\mathbb{L}(N, m_0, \delta_0)$ in Section 4.2.

Definition 4.4 *The complex input space $\mathbb{D}(N, m_0, \delta_0)$ is a set of sequences w expressed by*

$$w = \sum_{j=1}^{N} w^{(j)} \tag{4.3.14}$$

and satisfying that

$$[w^{(1)}, w^{(2)}, \ldots, w^{(N)}] \in \mathbb{D}(m_{01}, \delta_{01}) \times \mathbb{D}(m_{02}, \delta_{02}) \times \ldots \times \mathbb{D}(m_{0N}, \delta_{0N}). \tag{4.3.15}$$

Clearly, $\mathbb{D}(N, m_0, \delta_0)$ includes a transient input if one of elements in m_0 is infinity, and a persistent input if one of elements in m_0 is finite.

4.4 Performance Criterion

In this section, an appropriate performance measure for the standard discrete time systems $S_D(P, C)$, introduced in Section 2.3, is defined with their corresponding input spaces. This measure follows from the measure in Zakian [Zak86a].

Let $\mathbb{E}$ represent all input spaces in the discrete time domain.

Let $\bar{w} = [r, w]$. The output of the system $S_D(P, C)$

$$v(\bar{w}, C) : k \mapsto v(k, \bar{w}, C), \quad k \in \mathbb{N}, \tag{4.4.1}$$

depends upon the external input $\bar{w}$ and the controller C.

The measure of performance of the system $S_D(P, C)$ with external inputs in $\mathbb{E}$ is defined as

$$\phi_E(C) = \sup \{|v(k, \bar{w}, C)| : k \in \mathbb{N}, \bar{w} \in \mathbb{E}\}. \tag{4.4.2}$$

Roughly speaking, $\phi_E(C)$ is the largest absolute value of the generalised output v for all external inputs $\bar{w}$ in $\mathbb{E}$ as time k ranges over $\mathbb{N}$.

In critical control systems, the main objective is to ensure that during all time k, the generalised output remains tolerably small. A control system is said to be well behaved if it satisfies a performance criterion that includes, among other conditions, a mathematical statement of how small the performance function has to be. An appropriate mathematical statement for critical discrete control system design is expressed in the form of the following criterion:

$$\phi_E(C) \leq \varepsilon_d, \tag{4.4.3}$$

where ε_d is a finite positive number and is the largest tolerable value of $\phi_E(C)$.

The design criterion (4.4.3) gives rise to the problem of designing the controller C that satisfies the criterion. A significant difficulty encountered in solving this problem is that the evaluation of the numbers $\phi_E(C)$ is difficult if effected directly from the defining expression (4.4.2). This is mainly because of the supremum operation required over the input function space. To overcome this, some explicitly simple expressions for $\phi_E(C)$ are presented in the following chapters.

Chapter 5

Rejection of Persistent and/or Transient External Inputs

5.1 Introduction

The l^1-optimisation method [DP87b] can be used to design critical control systems with a certain class of input space (e.g. [PB90]). However, a critical control system is frequently subjected to persistent and/or transient external inputs and is often required to satisfy different performance criteria for different external inputs. In such cases, l^1-optimisation methods are limited. Thus, this chapter is devoted to the design of discrete systems with external inputs in the input spaces $\mathbb{L}(m, \delta)$ and/or $\mathbb{L}(N, m_0, \delta_0)$ defined in Section 4.2, including persistent and/or transient input spaces, to satisfy a set of performance criteria (or called a multi-performance criterion (MPC)) in the form used in the method of inequalities.

In Section 5.2, a multi-performance criterion which can cover some complicated control problems, each of which represents a performance requirement for a set of external inputs, is presented. Section 5.3 gives a form of a controller which satisfies the multi-performance criterion and some relations between an input space, a controller and output performance. In Section 5.4, the construction of the optimal solution for minimum phase plants with external inputs in persistent and/or transient input spaces, and non-minimum phase plants with external inputs in only persistent bounded input space, is developed. The numerical solution of the multi-performance problem and the determination of admissible controllers for non-minimum phase plants with persistent and/or transient external inputs are carried out by the method of inequalities in Section 5.5. Finally, in Section 5.6, the results of designing controllers by the method of inequalities are demonstrated by some examples.

5.2 Multi-Performance Criterion

In this section, a multi-performance criterion for systems with input spaces $\mathbb{L}(m, \delta)$ and/or $\mathbb{L}(N, m_0, \delta_0)$ is presented. It is shown that a particular case of the MPC is equivalent to the l^1-optimisation problem. The admissible controller for the MPC, and the optimal admissible controller, are defined.

Based on the performance criterion (4.4.3) along with the MI (1.5.4), the MPC is defined as

$$\phi_L(C, m_i) \leq \varepsilon_i, \quad i = 0, 1, 2, ..., M, \tag{5.2.1}$$

where $\phi_L(C, m_i)$ $(i = 1, 2, ..., M)$ represents the performance function $\phi_E(C)$ (4.4.2) of the system subjected to the external inputs in $\mathbb{L}(m_i, \delta_i)$, and ε_i $(i = 0, 1, ..., M)$ are finite positive numbers specified by the designers. In particular, $\phi_L(C, m_0)$ denotes the performance function $\phi_E(C)$ of the system with external inputs in $\mathbb{L}(N, m_0, \delta_0)$.

The performance criteria (5.2.1) are the general descriptions of the design problem for control systems. The number of performance criteria needed depends on the requirement of the system and the external environment (e.g. the external input types). For example, if the input $w(k) = 0.2(k+1)^{-1} \operatorname{sgn}(\beta(k))$, for $k < T_0$ and $w(k) = 0.1 \cos(k) \operatorname{sgn}(\alpha(k))$, for $k >> T_0$, where $\alpha(k)$ and $\beta(k)$ are random variables with zero mean, then the system may be required to satisfy the following criteria: $\phi_L(C, \infty) \leq \varepsilon_1$ and $\phi_L(C, 2) \leq \varepsilon_2$.

Consider a particular case of the multi-performance criterion. Assume that the external inputs are modelled by $\mathbb{L}(\infty, 1)$ and there is just one performance criterion, $\phi_L(C, \infty) \leq \varepsilon_1$. Letting

$$\varepsilon_1 = \min_C \{\phi_L(C, \infty)\} \tag{5.2.2}$$

is equivalent to the l^1-optimisation problem.

The performance criteria above also show that the system has different requirements for different input spaces. A controller satisfying the performance criteria (5.2.1) is called an admissible controller. The admissible controller that minimises all $\phi_L(C, m_i)$ $i = 0, 1, 2, ..., M$ is an optimal admissible controller, if such a controller exists.

5.3 Controller Structure and Output Performance

In this section, a form of a controller which satisfies the multi-performance criterion and some relations between an input space, a controller and output performance are developed.

Consider the plant described by the ARMAX model of the form

$$B_d y(k) = q^{-d} B_n u(k) + D_n \omega(k), \tag{5.3.1}$$

where the polynomials $B_d \in \mathbb{R}[q, \bar{n}]$ with $b_{d0} = 1$, $B_n \in \mathbb{R}[q, \bar{m}]$ and $D_n \in \mathbb{R}[q, \bar{l}]$ with $d_{n0} = 1$, the time delay $d \in \mathbb{N}^+ \backslash 0$, y, u and ω are the generalised output, the control and the external input of the system, respectively.

Assume that the polynomials B_d and $q^{-d}B_n$ are coprime. Then, by Propositions 2.2 and 2.4, there exist the following identities:

$$B_dF + q^{-d}E = D_n, \tag{5.3.2}$$
$$B_dQ + q^{-d}B_nP = D_n, \tag{5.3.3}$$

where the polynomials $F \in \mathbb{R}[q, d-1]$ with $f_0 = 1$, $E \in \mathbb{R}[q, \bar{n}-1]$, $Q \in \mathbb{R}[q, d+\bar{m}-1]$ with $q_0 = 1$ and $P \in \mathbb{R}[q, \bar{n}-1]$ are unique.

As previously stated, the objective is to design an admissible controller that will stabilise a given plant and satisfy the multi-performance criterion (5.2.1). Employing the famous YJB parameterisation of all stabilising controllers [YJB76], the admissible controller $C : -y \mapsto u$ has the following form.

$$C = \left(P + \bar{R}B_d\right)\left(Q - \bar{R}q^{-d}B_n\right)^{-1}, \quad \bar{R} \in \bar{A}_1, \tag{5.3.4}$$

where P and Q are determined by the identity (5.3.3).

By (5.3.1) and (5.3.4), then the relation between the generalised output y and the external input ω can be expressed by

$$y(k) = \left(Q - \bar{R}q^{-d}B_n\right)\omega(k). \tag{5.3.5}$$

Theorem 5.1 *Suppose the plant (5.3.1) is subjected to external inputs in* $\mathbb{L}(m, \delta)$. *Then the output performance is evaluated by*

$$\phi_L(C, m) = \left\| Q - \bar{R}q^{-d}B_n \right\|_{\bar{A}_n} \delta, \tag{5.3.6}$$

where $n^{-1} + m^{-1} = 1$.

Proof: Let $H = Q - \bar{R}q^{-d}B_n$. It follows from (5.3.5) that

$$y(k) = \sum_{i=0}^{k} h_i\omega(k-i). \tag{5.3.7}$$

Since $\omega \in \mathbb{L}(m, \delta)$ and $n^{-1} + m^{-1} = 1$, an application of Holder's inequality [Lue69] to the above and then a use of the definition of $\phi_L(C, m)$ give

$$\phi_L(C, m) \leq \left\| Q - \bar{R}q^{-d}B_n \right\|_{\bar{A}_n} \delta. \tag{5.3.8}$$

For every $k \in \mathbb{N}^+$, there is a sequence $\omega^* \in \mathbb{L}(m, \delta)$ defined by

$$\omega^*(k-i) = \delta\, |h_i|^{n-1} \, \| H \|_{\bar{A}_n}^{1-n} \operatorname{sgn}(h_i), \quad \text{for} \quad 0 \leq i \leq k \tag{5.3.9}$$

and $\omega^*(k-i) = 0$, otherwise.

Substituting (5.3.9) into (5.3.7) yields

$$y^*(k) = \sum_{i=0}^{k} |h_i|^n \, \| H \|_{\bar{A}_n}^{1-n} \delta. \tag{5.3.10}$$

As $k \to \infty$, then $y(\infty) = \| H \|_{\bar{A}_n} \delta$. Hence

$$\phi_L(C, m) \geq \left\| Q - \bar{R}q^{-d}B_n \right\|_{\bar{A}_n} \delta. \tag{5.3.11}$$

Therefore, the theorem follows from (5.3.8) and (5.3.11). □□

Theorem 5.1 can be extended simply to systems with complex inputs in $\mathbb{L}(N, m_0, \delta_0)$. We have the following corollary.

Corollary 5.1 *Suppose the plant (5.3.1) is subjected to external inputs in* $\mathbb{L}(N, m_0, \delta_0)$. *Then the output performance is*

$$\phi_L(C, m_0) = \sum_{i=1}^{N} \left\| Q - \bar{R}q^{-d}B_n \right\|_{\bar{A}_{n_{0i}}} \delta_{0i}, \tag{5.3.12}$$

where $n_{0i}^{-1} + m_{0i}^{-1} = 1$ $(i = 1, 2, ..., N)$.

The proof of the corollary is similar to that of Theorem 5.1; thus, the proof is omitted.

Suppose B_n is factorised as a product $B_s B$, where the zeros of B are all inside the close unit disc and the zeros of B_s are all outside the unit disc, and let $R = \bar{R}B_s$. Then $Q - \bar{R}q^{-d}B_n = Q - Rq^{-d}B$. From (5.3.2) and (5.3.3), we have $Q = F + Dq^{-d}$, where $D \in \mathbb{R}[q, \bar{m} - 1]$. Thus $Q - \bar{R}q^{-d}B_n = F + (D - RB)q^{-d}$ and

$$\left\| Q - \bar{R}q^{-d}B_n \right\|_{\bar{A}_n} = \left(\| F \|_{\bar{A}_n}^n + \| D - RB \|_{\bar{A}_n}^n \right)^{1/n}. \tag{5.3.13}$$

Therefore, it is clear that the minimisation of $\left\| Q - \bar{R}q^{-d}B_n \right\|_{\bar{A}_n}$ is converted into the minimisation of $\| D - RB \|_{\bar{A}_n}$, i.e

$$\min_{\bar{R} \in \bar{A}_1} \left\{ \left\| Q - \bar{R}q^{-d}B_n \right\|_{\bar{A}_n} \right\} \Longleftrightarrow \min_{R \in \bar{A}_1} \left\{ \| D - RB \|_{\bar{A}_n} \right\}. \tag{5.3.14}$$

From the above, the controller depends on the polynomial R. Now we replace $\phi_L(C, m_i)$ by $\phi_L(R, m_i)$. The design problem can then be stated as follows: given polynomials $D \in \mathbb{R}[q, \bar{m} - 1]$, $B \in \mathbb{R}[q, r]$, and $d \in \mathbb{N}^+ \backslash 0$, find an $R \in \bar{A}_1$ such that the performance criteria

$$\phi_L(R, m_i) \leq \varepsilon_i, \quad i = 0, 1, ..., M \tag{5.3.15}$$

are satisfied.

Denote R^o as a solution of $R \in \bar{A}_1$ which satisfies (5.3.15). Then the admissible controller is given by

$$C^o = (PB_s + R^o B_d) \left(\left(Q - R^o q^{-d} B \right) B_s \right)^{-1}. \tag{5.3.16}$$

Notice that since there may exist a set of $R \in \bar{A}_1$ satisfying (5.3.15), the admissible controller (5.3.16) is generally not unique.

5.4 Construction of the Optimal Solution

In this section, the minimisation of $\phi_L(R, m_i)$ $(i = 0, 1, 2, ..., M)$ for minimum phase plants is first considered. Then the minimisation of the performance function $\phi_L(R, \infty)$ of systems with only persistent external inputs in $\mathbb{L}(\infty, \delta)$ is developed in two cases. One is the special case where the plant (5.3.1) has only one zero inside the closed unit disc and has at least one zero inside the closed unit disc with an additional special condition. The other is the general case where there are no simplifying assumptions on the plant.

5.4.1 Minimum Phase Plants

For minimum phase plants, we have the following theorem.

Theorem 5.2 *Assume that the plant (5.3.1) is minimum phase. Then an optimal admissible controller is given by*

$$C^o = E(B_n F)^{-1} \tag{5.4.1}$$

and gives the optimal performance functions $\phi_L(C^o, m_i)$ which are minimised, i.e.

$$\phi_L(C^o, m_i) = \| F \|_{\bar{A}_{n_i}} \delta_i, \quad n_i^{-1} + m_i^{-1} = 1, \quad i = 1, 2, \ldots, M, \tag{5.4.2}$$

$$\phi_L(C^o, m_0) = \sum_{j=1}^{N} \| F \|_{\bar{A}_{n_{0i}}} \delta_{0i}, \quad n_{0j}^{-1} + m_{0j}^{-1} = 1, \quad j = 1, 2, \ldots, N. \tag{5.4.3}$$

Proof: Since the plant (5.3.1) is assumed to be minimum phase, $B_s = B_n$ and $B = 1$. Then $R^o = D$ achieves the minimum of $\| D - R \|_{\bar{A}_{n_i}}$ because $\| D - R^o \|_{\bar{A}_{n_i}} = 0$.

Substituting R^o into (5.3.16) and using (5.3.2) and (5.3.3) give the admissible controller C^o (5.4.1). As $R^o = D$, it follows from (5.3.13) that

$$\begin{aligned} \min_{\bar{R} \in \bar{A}_1} \left\{ \left\| Q - \bar{R} q^{-d} B_n \right\|_{\bar{A}_{n_i}} \right\} &= \left(\| F \|^{n_i}_{\bar{A}_{n_i}} + \| D - R^o B \|^{n_i}_{\bar{A}_{n_i}} \right)^{1/n_i} \\ &= \| F \|_{\bar{A}_{n_i}}, \quad i = 1, 2, \ldots, M. \end{aligned} \tag{5.4.4}$$

By Theorem 5.1, the optimal output performance (5.4.2) follows. Similarly, using Corollary 5.1 gives (5.4.3). □□

Example 5.1: Consider the plant

$$y(k) - 0.9y(k-1) = 0.5u(k-2) + \omega(k) + 0.7\omega(k-1). \tag{5.4.5}$$

Using the identity (5.3.2) gives $F = 1 + 1.6q^{-1}$ and $E = 1.44$. Thus, the optimal admissible controller C^o is

$$u(k) = -1.6u(k-1) - 2.88y(k) \tag{5.4.6}$$

and the optimal output performance is $\phi_L(C^o, 1) = 0.16$, for $\omega \in \mathbb{L}(1, 0.1)$; $\phi_L(C^o, 2) = 0.18868$, for $\omega \in \mathbb{L}(2, 0.1)$; $\phi_L(C^o, \infty) = 0.26$, for $\omega \in \mathbb{L}(\infty, 0.1)$; and $\phi_L(C^o, [2, \infty]) = 0.31868$, for $\omega \in \mathbb{L}(1, [2, \infty], [0.1, 0.05])$.

Note that the admissible controller C^o (5.4.1) is same as the minimum variance controller [Ast70].

5.4.2 Systems with Only Persistent External Inputs

We now study the minimisation of the output performance function of systems with only persistent external inputs in $\mathbb{L}(\infty, \delta)$. It is known from (5.3.13) that

$$\min_{\bar{R} \in \bar{A}_1} \left\{ \left\| Q - \bar{R} q^{-d} B_n \right\|_{\bar{A}_1} \right\} = \| F \|_{\bar{A}_1} + \min_{R \in \bar{A}_1} \left\{ \| D - RB \|_{\bar{A}_1} \right\}. \tag{5.4.7}$$

Thus, we only need to consider the minimisation of $\| D - RB \|_{\bar{A}_1}$. First, we deal with the special cases.

Theorem 5.3 *Suppose that B has only one zero within the closed unit disc at $q^{-1} = \sigma$ ($|\sigma| \le 1$). Then there exists an $R \in \mathbb{R}[q, \bar{m} - 2]$ such that $\| D - RB \|_{\bar{A}_1}$ is minimised,*

$$\min_{R \in \mathbb{R}[q, \bar{m}-2]} \left\{ \| D - RB \|_{\bar{A}_1} \right\} = |D(\sigma)| \tag{5.4.8}$$

and the unique choice of R^o that obtains the minimum is

$$R^o = \left(D(q^{-1}) - D(\sigma) \right) \left(q^{-1} - \sigma \right)^{-1}. \tag{5.4.9}$$

Proof: Since $\| D - RB \|_\infty \le \| D - RB \|_{\bar{A}_1}$ for each $R \in \bar{A}_1$, it follows that

$$\min_{R \in \bar{A}_1} \{ \| D - RB \|_\infty \} \le \min_{R \in \bar{A}_1} \left\{ \| D - RB \|_{\bar{A}_1} \right\}. \tag{5.4.10}$$

Now it is known [ZF83, FZ84, Vid86] that the choice of R^o in (5.4.9) achieves the minimum on the left side of (5.4.8). With the choice of R^o, $D - R^o B$ equals the constant function $D(\sigma)$, and $\| D - RB \|_\infty$ and $\| D - RB \|_{\bar{A}_1}$ are each equal to $D(\sigma)$. Thus, the result follows. □□

Example 5.2: Let the plant be

$$\begin{aligned} y(k) - 0.95 y(k-1) &= u(k-2) + 1.5 u(k-3) - u(k-4) \\ &\quad + \omega(k) - 0.7 \omega(k-1) \end{aligned} \tag{5.4.11}$$

with $\omega \in \mathbb{L}(\infty, 0.1)$.

By the identities (5.3.2) and (5.3.3), we have $F = 1 + 0.25 q^{-1}$, $E = 0.2375$, $Q = 1 + 0.25 q^{-1} - 0.076 q^{-2} - 0.1697 q^{-3}$ and $P = 0.1617$. Then $D = -0.076 - 0.1697 q^{-1}$ and $B = 0.5 + q^{-1}$. Thus, by Theorem 5.3, $R^o = -0.1697$, the optimal admissible controller C^o is

$$C^o = \frac{0.1537 - 0.00485 q^{-1}}{2 - 0.5 q^{-1} - 0.2323 q^{-2} - 0.0085 q^{-3}} \tag{5.4.12}$$

and the optimal output performance is $\phi_L(C^o, \infty) = 0.125885$.

Theorem 5.4 *Suppose that the polynomial $D \in \mathbb{R}[q, \bar{m}-1]$ and the coefficients of the unstable polynomial $B \in \mathbb{R}[q, r]$ satisfy*

$$|b_r| \geq \sum_{i=0}^{r-1} |b_i| \,. \tag{5.4.13}$$

Then there exists a polynomial $R \in \mathbb{R}[q, \bar{m}-r-1]$ such that

$$\min_{R \in \bar{A}_1} \left\{ \| D - RB \|_{\bar{A}_1} \right\} = \min_{R \in \mathbf{R}[q, \bar{m}-r-1]} \left\{ \| D - RB \|_{\bar{A}_1} \right\}. \tag{5.4.14}$$

Proof: Let $H = D - RB$. For $R \in \mathbb{R}[q, k]$ ($k \in \mathbb{N}^+$ and $k + r \geq \bar{m} - 1$), define

$$J_k = \sum_{i=0}^{r+k} |h_i| \,. \tag{5.4.15}$$

If $k = \bar{m} - r$, then it follows from the extension of $H = D - RB$ that

$$h_{\bar{m}} = b_r r_{\bar{m}-r}, \tag{5.4.16}$$

$$h_{\bar{m}-1} = d_{\bar{m}-1} - b_r r_{\bar{m}-r-1} - b_{r-1} r_{\bar{m}-r}, \tag{5.4.17}$$

$$h_{\bar{m}-2} = d_{\bar{m}-2} - b_r r_{\bar{m}-r-2} - b_{r-1} r_{\bar{m}-r-1} - b_{r-2} r_{\bar{m}-r}, \tag{5.4.18}$$

$$\cdots\cdots\cdots\cdots$$

It is clear from the above that

$$J_{\bar{m}-r} \geq J_{\bar{m}-r-1} + \left[|b_r| - \sum_{i=0}^{r-1} |b_i| \right] |r_{\bar{m}-r}| \tag{5.4.19}$$

which gives, by minimisation,

$$\min_{R \in \mathbf{R}[q, \bar{m}-r]} \{J_{\bar{m}-r}\} \geq \min_{R \in \mathbf{R}[q, \bar{m}-r-1]} \{J_{\bar{m}-r-1}\} + \min_{r_{\bar{m}-r} \in \mathbf{R}} \left\{ \left[|b_r| - \sum_{i=0}^{r-1} |b_i| \right] |r_{\bar{m}-r}| \right\}. \tag{5.4.20}$$

Hence, it follows from (5.4.13) and (5.4.20) immediately that $r_{\bar{m}-r} = 0$ and

$$\min_{R \in \mathbf{R}[q, \bar{m}-r]} \{J_{\bar{m}-r}\} \geq \min_{R \in \mathbf{R}[q, \bar{m}-r-1]} \{J_{\bar{m}-r-1}\} \tag{5.4.21}$$

which implies that the optimal R belongs in $\mathbb{R}[q, \bar{m}-r-1]$. Thus, the theorem is established. □□

Example 5.3: Let $D = 1.31 - 1.92q^{-1} + 0.78q^{-2} + 1.17q^{-3}$ and $B = 1 + 0.2q^{-1} - 1.43q^{-2}$. Since B satisfies (5.4.13), Theorem 5.4 shows that the optimal solution $R^o \in \mathbb{R}[q, 1]$. In fact, $R^o = -0.6599 - 0.8182q^{-1}$ and the minimum $\| D - R^o B \|_{\bar{A}_1} = 2.9397$.

Next, the problem of minimising $\| D - RB \|_{\bar{A}_1}$ is studied without any simplifying assumption on the polynomial B.

Theorem 5.5 *There exist an upper bound and a lower bound on the minimum of* $\| D - RB \|_{\bar{A}_1}$ *i.e.*

$$\min_{R \in \bar{A}_1} \{\| D - RB \|_\infty\} \le \min_{R \in \bar{A}_1} \left\{\| D - RB \|_{\bar{A}_1}\right\} \le \| D \|_{\bar{A}_1} \tag{5.4.22}$$

Proof: From the definitions of $\| . \|_\infty$ and $\| . \|_{\bar{A}_1}$, then

$$\min_{R \in \bar{A}_1} \{\| D - RB \|_\infty\} \le \min_{R \in \bar{A}_1} \left\{\| D - RB \|_{\bar{A}_1}\right\}. \tag{5.4.23}$$

Setting $R = 0$ gives $D - RB = D$. Hence

$$\min_{R \in \bar{A}_1} \left\{\| D - RB \|_{\bar{A}_1}\right\} \le \| D \|_{\bar{A}_1}. \tag{5.4.24}$$

The theorem follows from (5.4.23) and (5.4.24). □□

Theorem 5.5 shows that we can use an H^∞-optimisation method to obtain a lower bound of $\| D - RB \|_{\bar{A}_1}$. This is very useful in checking whether the performance criterion $\phi_L(R, \infty) \le \varepsilon_1$ is realistic.

Theorem 5.6 *Suppose that* $D \in \mathbb{R}[q, \bar{m} - 1]$ *and* $B \in \mathbb{R}[q, r]$ *with distinct unstable zeros. Then there exists a finite* $N_0 \in \mathbb{N}^+$ *such that*

$$\min_{R \in \bar{A}_1} \left\{\| D - RB \|_{\bar{A}_1}\right\} = \min_{R \in \mathbb{R}[q, N_0]} \left\{\| D - RB \|_{\bar{A}_1}\right\}. \tag{5.4.25}$$

Proof: Let $H = D - RB$. It is known [DP87b] that if $\| H \|_{\bar{A}_1}$ is minimised, then there exists a finite $M_0 \in \mathbb{N}^+$ such that the optimal $H^o \in \mathbb{R}[q, M_0]$ and satisfies

$$H^o(\sigma_i) = D(\sigma_i), \quad i = 1, 2, ..., r, \tag{5.4.26}$$

where σ_i is the zero of B, i.e. $B(\sigma_i) = 0$.

Then, $H^o - D$ can be factorised into $R^o B$, where $R^o \in \mathbb{R}[q, M_0 - r]$. Obviously, setting $R = R^o$ minimises $\| D - RB \|_{\bar{A}_1}$. Therefore, letting $N_0 = M_0 - r$ gives the result. □□

Example 5.4: Suppose that $D = 1.15 - 0.8q^{-1} + 1.75q^{-2} - 0.34q^{-3}$ and $B = 1 + 2.3q^{-1} + 1.32q^{-2}$. The optimal solution of $\min\{\| D - RB \|_{\bar{A}_1} : R \in \mathbb{R}[q, N_0]\}$ is $R^o = 1.15 - 0.2576q^{-1}$ and $\| D - R^o B \|_{\bar{A}_1} = 4.0118$, which are obtained by the numerical method discussed in the next section. Thus, $N_0 = 1$.

According to the theorems in this section, we can obtain directly the optimal $R^o \in \bar{A}_1$, which minimises $\| D - RB \|_{\bar{A}_1}$, by

$$\min_{R \in \mathbb{R}[q, N_0]} \left\{\| D - RB \|_{\bar{A}_1}\right\}. \tag{5.4.27}$$

Since the order of the polynomial $D - RB$ is $\max\{\bar{m} - 1, r + N_0\}$ and is finite, the minimisation of $\| D - RB \|_{\bar{A}_1}$ is converted into a simple problem of solving a function minimum, which can be easily solved by numerical methods.

Though we do not intend to discuss the construction of the optimal solution for systems with only one of other input spaces, e.g. $\mathbb{L}(2, \delta)$ or $\mathbb{L}(2, [2, \infty], [\delta_{01}, \delta_{02}])$, the idea studied in this section may be extended to those cases.

5.5 Numerical Solution of Admissible Controllers

In this section, the method of inequalities [ZAN73] is used to obtain the numerical solution of the admissible controller satisfying the performance criteria (5.2.1).

Let the vector $\mathbf{r}$ denote the coefficients of the polynomial $R \in \bar{A}_1$, i.e. $\mathbf{r} = [r_1, r_2, ...]$. Thus, we replace $\phi_L(R, m_i)$ by $\phi_L(\mathbf{r}, m_i)$ and the performance criteria (5.3.15) can be expressed as

$$\phi_L(\mathbf{r}, m_i) \leq \varepsilon_i, \quad i = 0, 1, 2, ..., M. \tag{5.5.1}$$

Even though l^1 and H^2 optimal design methods can obtain optimal controllers such that $\phi_L(\mathbf{r}, \infty)$ and $\phi_L(\mathbf{r}, 2)$ are minimised, respectively, it seems to be impossible for them to design an optimal controller satisfying the performance criteria above. Those methods are, in general, not applicable directly to the design of the admissible controller.

Let $\mathcal{S}_i$ be the set of parameter vectors for which the i-th performance criterion is satisfied:

$$\mathcal{S}_i = \{\mathbf{r} : \phi_L(\mathbf{r}, m_i) \leq \varepsilon_i\}. \tag{5.5.2}$$

Then an admissible or feasible set of parameter vectors for which all the performance criteria hold is the intersection

$$\mathcal{S} = \bigcap_{i=0}^{M} \mathcal{S}_i. \tag{5.5.3}$$

Zakian and Al-Naib [ZAN73] suggested an algorithm for solving this problem which they called the moving boundaries process (MBP), which is discussed in Section 3.2. Thus, we employ this algorithm in the design of the admissible controller.

First, set $\mathbf{r} \in \mathbb{R}^j$ $(j \in \mathbb{N}^+ \backslash 0)$ and use the MBP. If for $\mathbf{r} \in \mathbb{R}^j$, the set of performance criteria is not achievable, then on the basis of the result of the parameter vector $\mathbf{r} \in \mathbb{R}^j$, set $\mathbf{r} \in \mathbb{R}^{j+1}$ and repeat using the MBP until the set of criteria (5.5.1) is satisfied.

In case for $j > N_0$ (a finite number), one or some of performance criteria is still not satisfied and very little improvement of reduction of their performance function results from further increase in the dimension of the vector $\mathbf{r}$, it implies that the performance criteria may be inconsistent, while their magnitude gives some measure of how closely it is possible to approach the objectives. In this case, some of the performance criteria should be relaxed until they are satisfied. From the practical viewpoint, the approximate optimal solution is also appreciated if the optimal solution is almost impossible.

In addition, sometimes, the performance of critical systems mainly depends on one of the performance functions $\phi_L(\mathbf{r}, m_i)$ $(i = 0, 1, ..., M)$. This one is called a critical performance function. The other performance criteria can be taken as the constraints. In this case, we can also employ the moving boundaries process or a nonlinear programming algorithm (e.g. [McC83]) to obtain the optimal parameter vector $\mathbf{r}$ such that the critical performance function is minimised.

Note that although l^1, H^2 and H^∞-optimisation methods cannot be employed directly for the above problem, they can be used to check conditions that are necessary for a system to be designed. For example, if $\varepsilon_1 \leq \inf\{\phi_L(\mathbf{r}, 1)\}$, or $\varepsilon_2 \leq \inf\{\phi_L(\mathbf{r}, 2)\}$, then l^1, H^2 and H^∞-optimisation methods can determine whether or not an admissible controller for the system exists. If an admissible controller does not exist, then the design problem is not realistic and must be reformulated.

5.6 Examples

In this section, three examples are presented, which demonstrate the design of controllers satisfying certain performance criteria using the method of inequalities.

Example 5.5: Consider the plant

$$B_d = (1 + 0.5q^{-1})(1 - 0.7q^{-1})(1 + 0.9q^{-1}), \tag{5.6.1}$$
$$B_n = (1.2 - q^{-1})(1.5 + q^{-1})(0.9 + q^{-1})(0.8 + q^{-1}), \tag{5.6.2}$$
$$D_n = 1 + 0.9q^{-1}, \tag{5.6.3}$$
$$d = 2 \tag{5.6.4}$$

with external input $\omega \in \mathbb{L}(\infty, 0.05)$.

The objective is to design a controller C such that $\phi_L(C, \infty)$ is minimised. The polynomials B_d and B_n can be extended as

$$B_d = 1 + 0.7q^{-1} + 0.17q^{-2} + 0.35q^{-3}, \tag{5.6.5}$$
$$B_n = 1.296 + 2.844q^{-1} + 0.5q^{-2} - 2q^{-3} - q^{-4}. \tag{5.6.6}$$

Using the identities (5.3.2) and (5.3.3) gives

$$F = 1 + 0.2q^{-1}, \tag{5.6.7}$$
$$E = -0.31 - 0.315q^{-1} - 0.063q^{-2}, \tag{5.6.8}$$
$$Q = 1 + 0.2q^{-1} - 3.2464q^{-2} - 6.2599q^{-3} + 3.5333q^{-4} + 4.1799q^{-5}, \tag{5.6.9}$$
$$P = 3.9247 - 0.8098q^{-1} + 1.3167q^{-2}. \tag{5.6.10}$$

Thus, the polynomials D and B are given by

$$D = -3.2469 - 6.2599q^{-1} + 3.5333q^{-2} + 4.1799q^{-3}, \tag{5.6.11}$$
$$B = 0.72 + 1.7q^{-1} + q^{-2}. \tag{5.6.12}$$

Setting $\phi_L(C, \infty) \leq 0$ and using the method of inequalities give the optimal solution

$$R^o = -4.5089 + 4.1799q^{-1} \tag{5.6.13}$$

and the optimal output performance

$$\phi_L(\mathbf{r}^o, \infty) = 0.187035. \tag{5.6.14}$$

By (5.3.16), the optimal admissible controller C^o is given by

$$C^o = \frac{2.5556 - 1.6114q^{-1} + 0.8477q^{-2} - 0.2949q^{-3}}{1.8 + 0.06q^{-1} - 1.06q^{-2} - 3.0877q^{-3} + 2.1667q^{-4} + 1.3234q^{-5} - 0.9364q^{-6}}. \tag{5.6.15}$$

Example 5.6: In Example 5.5, suppose that the external input $\omega \in \mathbb{L}(2, 0.1)$ and the optimal solution of minimising $\phi_L(\mathbf{r}, 2)$ may belong in $\mathbb{R}[q, 2], \mathbb{R}[q, 3]$ or $\mathbb{R}[q, 4]$. Then, the results of the numerical solution are that

$$R_2^o = -5.5218 + 5.1793q^{-1} - 0.3445q^{-2}, \tag{5.6.16}$$
$$R_3^o = -5.5930 + 5.3398q^{-1} - 0.5601q^{-2} + 0.1821q^{-3}, \tag{5.6.17}$$
$$R_4^o = -5.6224 + 5.4103q^{-1} - 0.6664q^{-2} + 0.3015q^{-3} - 0.0912q^{-4} \tag{5.6.18}$$

and

$$\phi_L(\mathbf{r}_2^o, 2) = 0.15724, \tag{5.6.19}$$
$$\phi_L(\mathbf{r}_3^o, 2) = 0.15556, \tag{5.6.20}$$
$$\phi_L(\mathbf{r}_4^o, 2) = 0.15518. \tag{5.6.21}$$

Now, consider external inputs $\omega \in \mathbb{L}(2, [\infty, 2], [0.025, 0.06])$. Similarly, the results of the numerical solution of minimising $\phi_L(\mathbf{r}, [\infty, 2])$ for $R \in \mathbb{R}[q, 2]$ and $\mathbb{R}[q, 3]$ are that

$$R_2^o = -5.5043 + 5.1532q^{-1} - 0.3271q^{-2}, \tag{5.6.22}$$
$$R_3^o = -5.5647 + 5.2890q^{-1} - 0.5093q^{-2} + 0.1537q^{-3} \tag{5.6.23}$$

and

$$\phi_L(\mathbf{r}_2^o, [\infty, 2]) = 0.1890, \tag{5.6.24}$$
$$\phi_L(\mathbf{r}_3^o, [\infty, 2]) = 0.1883. \tag{5.6.25}$$

It can be seen from the above that R_2^o achieves significant reduction of performance function $\phi_L(\mathbf{r}, 2)$ and $\phi_L(\mathbf{r}, [\infty, 2])$, and little improvement results from further increase in the order of the polynomial R. This phenomenon almost always occurs although it has not yet been proved except with the performance function $\phi_L(\mathbf{r},\infty)$.

Example 5.7: Assume that the performance criteria of the system in Example 5.5 are

$$\phi_L(C,\infty) \le 0.1892, \quad \text{for} \quad \omega \in \mathbb{L}(\infty, 0.05), \tag{5.6.26}$$
$$\phi_L(C,2) \le 0.1892, \quad \text{for} \quad \omega \in \mathbb{L}(2, 0.12), \tag{5.6.27}$$
$$\phi_L(C,(\infty,2)) \le 0.1892, \quad \text{for} \quad \omega \in \mathbb{L}(2,[\infty,2],[0.025,006]). \tag{5.6.28}$$

Using the method of inequalities gives a solution

$$R^o = -5.4661 + 5.0870q^{-1} - 0.2631q^{-2} \tag{5.6.29}$$

and the output performance

$$\phi_L(\mathbf{r}^o,\infty) = 0.1892, \tag{5.6.30}$$
$$\phi_L(\mathbf{r}^o,2) = 0.1892, \tag{5.6.31}$$
$$\phi_L(\mathbf{r}^o,[\infty,2]) = 0.1892. \tag{5.6.32}$$

Next, assume that the design objective is to minimise the critical performance function $\phi_L(C,[\infty,2])$ for $\omega \in \mathbb{L}(2,[\infty,2],[0.025,006])$ subject to

$$\phi_L(C,\infty) \le 0.1892, \quad \text{for} \quad \omega \in \mathbb{L}(\infty, 0.05), \tag{5.6.33}$$
$$\phi_L(C,2) \le 0.1892, \quad \text{for} \quad \omega \in \mathbb{L}(2, 0.12). \tag{5.6.34}$$

Set the performance function criterion to $\phi_L(C,[\infty,2]) \le 0$. It is clear that there does not exist a solution for $\phi_L(C,[\infty,2]) \le 0$, but, in this way, we can find a minimum of $\phi_L(C,[\infty,2])$ and a solution R^o using the method of inequalities, which are

$$R^o = -5.4673 + 5.0981q^{-1} - 0.2895q^{-2} \tag{5.6.35}$$

and the output performance

$$\phi_L(\mathbf{r}^o,\infty) = 0.1892, \tag{5.6.36}$$
$$\phi_L(\mathbf{r}^o,2) = 0.1890, \tag{5.6.37}$$
$$\phi_L(\mathbf{r}^o,[\infty,2]) = 0.1891. \tag{5.6.38}$$

Chapter 6

Sup Controllers and Output Performance

6.1 Introduction

This chapter deals with the problem of controlling discrete critical systems with the input space $\mathbb{D}(m, \delta)$ or the complex input space $\mathbb{D}(N, m_0, \delta_0)$ defined in Section 4.3, starting with the design problem of minimum phase plants and concluding with the design problem of non-minimum phase plants. In the design of a controller for these systems one should consider the three important aspects of:

- Input space: This is concerned with environmental conditions which the system is subjected to, including the reference inputs and the disturbances.
- Output performance: This is concerned with the ability of the system to reach satisfactory outputs.
- Stability: This is concerned with stability of the system, including boundness of outputs.

In the chapter, a controller which minimises the performance function (3.4.2), called the sup controller, is developed for the systems with the input space $\mathbb{D}(m, \delta)$ or the complex input space $\mathbb{D}(N, m_0, \delta_0)$, which contains persistent inputs and/or transient inputs, and with a minimum phase plant. If there is no reference input, the sup controller is called the sup regulator since the design of the system is a regulation problem. Some relations are found between the input space, the sup controller and the optimal output performance. For non-minimum phase plants, the design of sup controllers can be resolved via numerical methods based on linear programming and the extrema of functions.

This chapter is arranged in the following way. In Section 6.2, a sup regulator for systems without reference input is presented, and it will be shown that the optimal output performance is related to the input space (here called the disturbance space) and the sup regulator. In Section 6.3, the sup controller is developed and some relations between the input space, the sup controller and the optimal output performance are given in simple formulae. Section 6.4 examines the stability of the closed-loop systems with the sup controllers. Section 6.5 establishes the uniqueness of the control generated by the sup controllers. Finally, sup controllers for the system $S_D(P,C)$ with a non-minimum phase plant are discussed in Section 6.6.

6.2 Sup Regulators and Output Performance

This section considers the system $S_D(P,C)$ without the reference input r. In this instance, the generalised output v is replaced by the output y, and the performance function (3.4.2) for the system $S_D(P,C)$ is replaced by

$$\phi_{DR}(C) = \sup\{|y(k,\omega,C)| : k \in \mathbb{N}, \omega \in \mathbb{D}\}, \tag{6.2.1}$$

where the input space $\mathbb{D}$ represents either $\mathbb{D}(m,\delta)$ or $\mathbb{D}(N,m_0,\delta_0)$.

This will then lead us to consider it as a regulation problem of control systems. The design problem is to find a regulator $C : y \mapsto u$ such that the performance function $\phi_{DR}(C)$ is minimised. Such a regulator is called a sup regulator.

For the sake of simplicity, let $D_n = 1$ in the ARMAX model describing the plant in the system $S_D(P,C)$. Then the plant is of the form

$$B_d y(k) = q^{-d} B_n u(k) + \omega(k), \tag{6.2.2}$$

where the polynomials $B_d \in \mathbb{R}[q,\bar{n}]$ with $b_{d0} = 1$ and $B_n \in \mathbb{R}[q,\bar{m}]$, the time delay $d \in \mathbb{N}^+ \backslash 0$, y, u and ω are the output, the control and the disturbance of the system, respectively.

By Proposition 2.3, there exists the following identity:

$$\Delta B_d F + q^{-d} E = 1, \tag{6.2.3}$$

where the polynomials $F \in \mathbb{R}[q,d-1]$ with $f_0 = 1$ and $E \in \mathbb{R}[q,\bar{n}]$ are unique.

Lemma 6.1 *The plant (6.2.2) can be expressed in a predictor form as*

$$y(k+d) = \xi(k) + \Delta F\omega(k+d), \tag{6.2.4}$$
$$\xi(k) = Ey(k) + \Delta F B_n u(k), \tag{6.2.5}$$

where the polynomials F and E are determined by the identity (6.2.3).

Proof: Multiplying (6.2.2) by $\Delta F q^d$ yields

$$\Delta B_d F y(k+d) = \Delta F B_n u(k) + \Delta F\omega(k+d). \tag{6.2.6}$$

From identity (6.2.3), the lemma follows. □□

Remark 6.1 *Lemma 6.1 shows that $y(k+d)$ consists of two terms: one depending on the past known controls together with measured outputs and the other depending on the disturbances. As F is of order $d-1$, the disturbance components are all in the future, so it is impossible to know the value of $\Delta F\omega(k+d)$ from the given measured output y and the control u up to time k without further dynamic modelling of the disturbance.*

Now, let us consider the input space $\mathbb{D} = \mathbb{D}(m,\delta)$. Since different values of m in the input space $\mathbb{D}(m,\delta)$ might lead to different output performance, two cases are studied.

6.2.1 Case 1: $0 \le m < d-1$

Let $|\Delta\omega(i)| = \gamma_i\delta$ for $i \in \mathbb{N}$, where $\gamma_i \ge 0$ ($i \in \mathbb{N}$). For $\omega \in \mathbb{D}(m,\delta)$, then

$$\sum_{i=k}^{k+m} \gamma_i \le 1, \quad \forall k \in \mathbb{N}. \tag{6.2.7}$$

By Lemma 6.1, we have

$$|y(k+d)| \le |\xi(k)| + \chi\left(\gamma_{k+1}, \gamma_{k+2}, \ldots, \gamma_{k+d}\right)\delta, \tag{6.2.8}$$

where

$$\chi\left(\gamma_{k+1}, \gamma_{k+2}, \ldots, \gamma_{k+d}\right) = \sum_{i=0}^{d-1} |f_i|\, \gamma_{k+d-i}. \tag{6.2.9}$$

From (6.2.8), the problem is to find a vector $[\gamma_{k+1}, \gamma_{k+2}, \ldots, \gamma_{k+d}]$ which maximises χ such that it maximises the linear form (i.e. the objective function)

$$\sum_{i=0}^{d-1} |f_i|\, \gamma_{k+d-i} \tag{6.2.10}$$

subject to the linear constraints

$$\gamma_{k+i} \ge 0, \quad i = 1, 2 \ldots, d \tag{6.2.11}$$

and, from (6.2.7),

$$\sum_{i=j}^{j+m} \gamma_i \le 1, \quad j = k+1, k+2, \ldots, k+d-m. \tag{6.2.12}$$

In fact, the above problem is a special case of the general linear programming problem. Using the simplex method [Coh78, Chv83], the optimal solution $[\gamma^o_{k+1}, \gamma^o_{k+2}, \ldots, \gamma^o_{k+d}]$ for the above problem can be obtained in a finite number of simplex stages. Hence

$$\max\left\{\chi\left(\gamma_{k+1}, \gamma_{k+2}, \ldots, \gamma_{k+d}\right)\right\} = \sum_{i=0}^{d-1} |f_i|\, \gamma^o_{k+d-i}. \tag{6.2.13}$$

From the above, we can see that as k ranges over $\mathbb{N}$, the linear programming problem above is not changed. So, all optimal solutions $[\gamma_{k+1}^o, \gamma_{k+2}^o, \ldots, \gamma_{k+d}^o]$ for $k \in \mathbb{N}$ are the same. Let $[\gamma_1^o, \gamma_2^o, \ldots, \gamma_d^o]$ denote these optimal solutions. Thus

$$\max\left\{\chi\left(\gamma_{k+1}, \gamma_{k+2}, \ldots, \gamma_{k+d}\right) : k \in \mathbb{N}\right\} = \sum_{i=0}^{d-1} |f_i|\, \gamma_{d-i}^o. \tag{6.2.14}$$

From (6.2.8), then

$$|y(k+d)| \le |\xi(k)| + \sum_{i=0}^{d-1} |f_i|\, \gamma_{d-i}^o \delta \tag{6.2.15}$$

which implies, by (6.2.1),

$$\phi_{DR}(C) \le \sup\left\{|\xi(k)| : k \in \mathbb{N}\right\} + \sum_{i=0}^{d-1} |f_i|\, \gamma_{d-i}^o \delta. \tag{6.2.16}$$

Choose a special disturbance $\omega^* \in \mathbb{D}(m, \delta)$ defined by

$$\omega^*(k) = \begin{cases} 0, & k \le 0 \\ \omega^*(k-1) + \delta\gamma_k^o \operatorname{sgn}\left(f_{d-k}\right), & 0 < k \le d \\ \omega^*(k-1), & k > d \end{cases} \tag{6.2.17}$$

Consider also a regulator C_R defined by

$$\Delta F B_n u(k) = -E y(k). \tag{6.2.18}$$

Since this implies that $\xi(k) = 0$, it follows from (6.2.4) that

$$y(d, \omega^*, C_R) = \sum_{i=0}^{d-1} |f_i|\, \gamma_{d-i}^o \delta. \tag{6.2.19}$$

With Remark 6.1, it is clear from (6.2.16) and (6.2.19) that $\phi_{DR}(C)$ cannot be smaller than $y(d, \omega^*, C_R)$. Therefore, if a regulator C can be found that makes

$$\phi_{DR}(C) = \sum_{i=0}^{d-1} |f_i|\, \gamma_{d-i}^o \delta, \tag{6.2.20}$$

then the regulator minimises $\phi_{DR}(C)$. In fact, the regulator C_R defined by (6.2.18) together with (6.2.16) and (6.2.19) implies that the minimum of the performance function is

$$\phi_{DR}(C_R) = \sum_{i=0}^{d-1} |f_i|\, \gamma_{d-i}^o \delta \tag{6.2.21}$$

and the output is

$$y(k, w, C_R) = \Delta F \omega(k). \tag{6.2.22}$$

Summarising the above results gives the following theorem.

Theorem 6.1 *Consider the system $S_D(P, C)$ with the plant (6.2.2) and the input space $\mathbb{D}(m, \delta)$. Then for $0 \leq m < d-1$ in $\mathbb{D}(m, \delta)$, the sup regulator C_R defined by (6.2.18) minimises $\phi_{DR}(C)$, and gives the optimal performance (6.2.21) and the output (6.2.22).*

Example 6.1: Let us consider the plant

$$y(k) + 0.7y(k-1) = 0.5u(k-3) + \omega(k) \tag{6.2.23}$$

with the disturbance $\omega \in \mathbb{D}(1, 0.1)$.

By the identity (6.2.3), the polynomials F and E are, respectively,

$$F = 1 + 0.3q^{-1} + 0.79q^{-2}, \tag{6.2.24}$$

$$E = 0.447 + 0.553q^{-1}. \tag{6.2.25}$$

From (6.2.18), the sup regulator C_R is

$$u(k) = -0.894y(k) - 1.106y(k-1) + 0.7u(k-1) - 0.49u(k-2) + 0.79u(k-3). \tag{6.2.26}$$

In this case, the objective function to be minimised for the linear programming is $\gamma_1 + 0.3\gamma_2 + 0.79\gamma_3$ subject to the constraints $\gamma_i \geq 0$, for $i = 1, 2, 3$, $\gamma_1 + \gamma_2 \leq 1$ and $\gamma_2 + \gamma_3 \leq 1$. It is easy to see that the optimal solution of the linear programming is $[\gamma_1^o, \gamma_2^o, \gamma_3^o] = [1, 0, 1]$. Therefore,

$$\phi_{DR}(C_R) = (1 + 0.79) \times 0.1 = 0.179. \tag{6.2.27}$$

Remark 6.2 *For the special case $m = 0$, we can easily see that the optimal solution $[\gamma_1^o, \gamma_2^o, \ldots, \gamma_d^o]$ is $[1, 1, ..., 1]$. In fact, in this case, the linear programming problem above can simply be stated as the objective function*

$$\sum_{i=0}^{d-1} |f_i| \, \gamma_{d-i}^o \tag{6.2.28}$$

subject to the linear constraints

$$0 \leq \gamma_i \leq 1, \quad i = 1, 2..., d. \tag{6.2.29}$$

Obviously, the optimal solution of the problem is $\gamma_i = 1$ for $i = 1, 2, ..., d$. Therefore, for $m = 0$,

$$\phi_{DR}(C_R) = \sum_{i=0}^{d-1} |f_i| \, \delta. \tag{6.2.30}$$

In Example 6.1, if the disturbance $\omega \in \mathbb{D}(0, 0.1)$, then the optimal output performance:

$$\phi_{DR}(C_R) = (1 + 0.3 + 0.79) \times 0.1 = 0.209. \tag{6.2.31}$$

6.2.2 Case 2: $m \geq d-1$

In this case, the following theorem is obtained.

Theorem 6.2 *If $m \geq d-1$ in $\mathbb{D}(m, \delta)$, then the sup regulator C_R that minimises $\phi_{DR}(C)$ is given by (6.2.18) and gives that*

$$\phi_{DR}(C_R) = \| F \|_{\bar{A}_\infty} \delta \tag{6.2.32}$$

and (6.2.22) holds.

Proof: It follows from (6.2.4) that

$$\begin{aligned} |y(k+d)| &\leq |\xi(k)| + |f_0|\,|\Delta\omega(k+d)| + |f_1|\,|\Delta\omega(k+d-1)| + \ldots \\ &\quad + |f_{d-1}|\,|\Delta\omega(k+1)| \\ &\leq |\xi(k)| + \| F \|_{\bar{A}_\infty} \sum_{i=k+1}^{k+d} |\Delta\omega(i)|\,. \end{aligned} \tag{6.2.33}$$

where f_i $(i = 0, 1, ..., d-1)$ is the coefficient of the polynomial F.

Since $\omega \in \mathbb{D}(m, \delta)$ and $m \geq d-1$, we have

$$|y(k+d)| \leq |\xi(k)| + \| F \|_{\bar{A}_\infty} \delta. \tag{6.2.34}$$

By (6.2.1), it can be deduced that

$$\phi_{DR}(C) \leq \sup\{|\xi(k)| : k \in \mathbb{N}\} + \| F \|_{\bar{A}_\infty} \delta. \tag{6.2.35}$$

Consider a particular disturbance ω^* defined by

$$\Delta\omega^*(k) = \begin{cases} \delta \text{ sgn } (f_M), & k = M \\ 0, & k \neq M \end{cases} \tag{6.2.36}$$

where f_M belongs to $\{f_0, f_1, \cdots, f_{d-1}\}$ and has the greatest absolute value in $\{f_0, f_1, \cdots, f_{d-1}\}$. Clearly, $\omega^* \in \mathbb{D}(m, \delta)$. Consider also the regulator C_R defined by (6.2.18). Since this implies that $\xi(k) = 0$, it follows that

$$y(d, \omega^*, C_R) = \delta\,|f_M| = \| F \|_{\bar{A}_\infty} \delta. \tag{6.2.37}$$

With the same reasoning as in Case 1, the theorem follows from (6.2.35) and (6.2.37) immediately. □□

In Example 6.1, assume the disturbance $\omega \in \mathbb{D}(2, 0.1)$. Then the minimum of the performance function is $\phi_{DR}(C_R) = 0.1$.

Note that from two cases above, it is clear that the sup regulator C_R is independent of the parameters m and δ of the input space $\mathbb{D}(m, \delta)$.

Define a function

$$Q_R(k) = \begin{cases} \sum_{i=0}^{d-1} |f_i| \gamma^o_{d-i}, & 0 \le k < d-1 \\ \| F \|_{\bar{A}_\infty}, & k \ge d-1 \end{cases} \tag{6.2.38}$$

where γ_i^o $(i = 1, 2, .., d)$ is the same as that of Case 1.

A generalisation of Theorems 1 and 2 is obtained for systems with the input space $\mathbb{D} = \mathbb{D}(N, m_0, \delta_0)$ as follows.

Theorem 6.3 *Suppose that the disturbance $\omega \in \mathbb{D}(N, m_0, \delta_0)$. Then the sup regulator C_R defined by (6.2.18) minimises $\phi_{DR}(C)$ and leads to*

$$\phi_{DR}(C_R) = \sum_{i=1}^{N} Q_R(m_{0i})\delta_{0i}, \tag{6.2.39}$$

$$y(k, \omega, C_R) = F \sum_{i=1}^{N} \Delta\omega^{(i)}(k), \tag{6.2.40}$$

where the polynomial F is the same as that in (6.2.3).

Proof: Immediate from the combination of Theorems 1 and 2. □□

Assume that the disturbance $\omega \in \mathbb{D}(3, [0, 1, 2], [0.02, 0.03, 0.05])$ in Example 6.1. Then, by Theorem 6.3, $\phi_{DR}(C_R) = 2.09 \times 0.02 + 1.79 \times 0.03 + 1 \times 0.05 = 0.1455$.

Summarising the results of Theorems 6.1-6.3 results in

$$\phi_{DR}(C_R) = \begin{cases} Q_R(m)\delta, & \mathbb{D} = \mathbb{D}(m, \delta) \\ \sum_{i=1}^{N} Q_R(m_{0i})\delta_{0i}, & \mathbb{D} = \mathbb{D}(N, m_0, \delta_0) \end{cases} \tag{6.2.41}$$

which provides the formulae for computing the optimal performance function and displays some simple relations between the input space $\mathbb{D}$, the sup regulator C_R and the optimal performance $\phi_{DR}(C_R)$.

6.3 Sup Controllers and Output Performance

This section is concerned with the problem of controlling the error $e = r - y$ of the system $S_D(P, C)$ with $\bar{w} \in \mathbb{D}_1 \times \mathbb{D}_2$ such that it is minimised as the reference input r ranges over $\mathbb{D}_1$, the disturbance ω ranges over $\mathbb{D}_2$ and time t ranges over $\mathbb{N}$, where $\mathbb{D}_1$ denotes either $\mathbb{D}(m^{(1)}, \delta^{(1)})$ or $\mathbb{D}(N, m_0^{(1)}, \delta_0^{(1)})$ and in a similar way $\mathbb{D}_2$ denotes either $\mathbb{D}(m^{(2)}, \delta^{(2)})$ or $\mathbb{D}(N, m_0^{(2)}, \delta_0^{(2)})$. Then, following (3.4.2), the performance function is defined as

$$\phi_D(C) = \sup\left\{|e(k, \bar{w}, C)| : k \in \mathbb{N}, \bar{w} \in \mathbb{D}_1 \times \mathbb{D}_2\right\}, \tag{6.3.1}$$

where the external input vector $\bar{w} = [r, \omega]$.

Let the plant description of the system $S_D(P, C)$ be the same as (6.2.2), i.e.

$$B_d y(k) = q^{-d} B_n u(k) + \omega(k). \tag{6.3.2}$$

To ease the presentation we first consider the special case in which the disturbance $\omega = 0$, so that the plant (6.3.2) becomes the ARMA model:

$$B_d y(k) = q^{-d} B_n u(k). \tag{6.3.3}$$

The performance function (6.3.1) is simplified to

$$\phi_{DC}(C) = \sup \{|e(k, r, C)| : k \in \mathbb{N}, r \in \mathbb{D}_1\}. \tag{6.3.4}$$

Define a function

$$Q_C(k) = \begin{cases} \sum_{i=0}^{d-1} \gamma_i^o, & 0 \le k < d-1 \\ 1, & k \ge d-1 \end{cases} \tag{6.3.5}$$

where γ_i^o $(i = 0, 1, ..., d-1)$ is the same as that of Section 6.2.

Theorem 6.4 *For the system $S_D(P, C)$ with the plant (6.3.3) and the reference input $r \in \mathbb{D}_1$, then the sup controller C_O, which minimises $\phi_{DC}(C)$, is given by*

$$\Delta F B_n u(k) = r(k) - E y(k), \tag{6.3.6}$$

and gives the minimum of the performance function

$$\phi_{DC}(C_O) = \begin{cases} Q_C(m^{(1)})\delta^{(1)}, & \mathbb{D}_1 = \mathbb{D}(m^{(1)}, \delta^{(1)}) \\ \sum_{i=1}^{N} Q_C(m_{0i}^{(1)})\delta_{0i}^{(1)}, & \mathbb{D}_1 = \mathbb{D}(N^{(1)}, m_0^{(1)}, \delta_0^{(1)}) \end{cases} \tag{6.3.7}$$

and the error

$$e(k, r, C_O) = r(k) - r(k-d). \tag{6.3.8}$$

Proof: Since the disturbance $\omega = 0$, it follows from Lemma 6.1 that

$$y(k+d) = \xi(k). \tag{6.3.9}$$

Subtracting $r(k+d)$ from both sides of (6.3.9) gives

$$\begin{aligned} y(k+d) - r(k+d) &= \xi(k) - r(k+d) \\ &= \xi(k) - r(k) - (r(k+d) - r(k)). \end{aligned} \tag{6.3.10}$$

Since $e(k) = r(k) - y(k)$,

$$e(k+d) = r(k) - \xi(k) + (r(k+d) - r(k)). \tag{6.3.11}$$

Let $r(k)-\xi(k)$ and $r(k+d)-r(k)$ replace $\xi(k)$ and $\Delta F\omega(k)$ in Cases 1 and 2 of Section 6.2, respectively. Then using the same reasoning as Cases 1 and 2, the results of the theorem follow immediately. □□

Note that the sup controller (6.3.6) is directly obtained by setting $\xi(k) = r(k)$.

Next, let us consider the general case in which there are both the reference input r and the disturbance ω.

Theorem 6.5 *Suppose that $\bar{w} \in \mathbb{D}_1 \times \mathbb{D}_2$ in the system $S_D(P, C)$. Then the sup controller C_O given by (6.3.6) minimises $\phi_D(C)$ and leads to*

$$\phi_D(C_O) = \phi_{DC}(C_O)|_{\mathbf{D}_1} + \phi_{DR}(C_R)|_{\mathbf{D}_2} \tag{6.3.12}$$

and the minimal error

$$e(k, \bar{w}, C_O) = \sum_{i=1}^{d} \Delta q^{-i} \left(r(k) - f_{d-i}\omega(k)\right), \tag{6.3.13}$$

where $(\cdot)|_{\mathbf{D}_j}$ $(j = 1, 2)$ means the condition that the input space is $\mathbb{D}_j$, and $\phi_{DC}(C_O)$ and $\phi_{DR}(C_R)$ are given by (6.3.7) and (6.2.41), respectively.

Proof: From Lemma 6.1 and since $e(k) = r(k) - y(k)$,

$$e(k, \bar{w}, C) = r(k) - \xi(k) + (r(k+d) - r(k)) - \Delta F\omega(k+d). \tag{6.3.14}$$

From the previous results, it follows that

$$\sup\{|r(k+d) - r(k)| : k \in \mathbb{N}, r \in \mathbb{D}_1\} \leq \phi_{DC}(C_O)|_{\mathbf{D}_1}, \tag{6.3.15}$$
$$\sup\{|\Delta F\omega(k+d)| : k \in \mathbb{N}, \omega \in \mathbb{D}_2\} \leq \phi_{DR}(C_R)|_{\mathbf{D}_2}. \tag{6.3.16}$$

which gives, with (6.3.1) and (6.3.14),

$$\phi_D(C_O) \leq \sup\{|r(k) - \xi(k)| : k \in \mathbb{N}\} + \phi_{DC}(C_O)|_{\mathbf{D}_1} + \phi_{DR}(C_R)|_{\mathbf{D}_2}. \tag{6.3.17}$$

According to the manner of choosing the particular disturbance in Section 6.2, it is easy to find a particular reference input r^* in $\mathbb{D}_1$ and a particular disturbance ω^* in $\mathbb{D}_2$ such that

$$r^*(k+d) - r^*(k) = \phi_{DC}(C_O)|_{\mathbf{D}_1}, \tag{6.3.18}$$
$$-\Delta F\omega^*(k+d) = \phi_{DR}(C_R)|_{\mathbf{D}_2}. \tag{6.3.19}$$

Consider the sup controller C_O given by (6.3.6). Since it implies $\xi(k) = r(k)$, it follows from (6.3.18) and (6.3.19) that

$$e(k, \bar{w}^*, C_O) = \phi_{DC}(C_O)|_{\mathbf{D}_1} + \phi_{DR}(C_R)|_{\mathbf{D}_2}. \tag{6.3.20}$$

With (6.3.17) and (6.3.20), it can be deduced that

$$\phi_D(C_O) = \phi_{DC}(C_O)|_{\mathbf{D}_1} + \phi_{DR}(C_R)|_{\mathbf{D}_2}. \tag{6.3.21}$$

Substituting $\xi(k) = r(k)$ into (6.3.14) gives

$$\begin{aligned} e(k, \bar{w}, C_O) &= r(k) - r(k-d) - \Delta F\omega(k) &(6.3.22)\\ &= \sum_{i=1}^{d} \Delta q^{-i} \left(r(k) - f_{d-i}\omega(k)\right). &(6.3.23)\end{aligned}$$

Therefore, the theorem is proved. □□

Let us see an example which illustrates the properties of the sup controller.

Example 6.2: Let the plant be

$$y(k) + 0.5y(k-1) + 0.1y(k-2) = u(k-3) + 0.8u(k-4) + \omega(k). \tag{6.3.24}$$

The reference input r and the disturbance ω are chosen as

$$r(k+1) = r(k) + \begin{cases} 0.03, & k \in [0, 200) \\ 0, & k \in [200, 800) \\ -0.03, & k \in [800, \infty) \end{cases} + 0.02 \text{ sgn } (\alpha(k)), \tag{6.3.25}$$

$$w(k+1) = w(k) + 0.08 \text{ sgn } (\beta(k)), \tag{6.3.26}$$

where $\alpha(k)$ and $\beta(k)$ are random variables ranged over $[-1, 1]$ with zero mean and with variance equal to one. Clearly, the external input vector $[r, \omega] \in \mathbb{D}(0, 0.05) \times \mathbb{D}(0, 0.08)$.

By Theorem 6.4, the sup controller for the plant is

$$\begin{aligned} u(k) &= r(k) - 0.625y(k) - 0.31y(k-1) - 0.065y(k-2) - 0.3u(k-1) + \\ &\quad +0.25u(k-2) + 0.53u(k-3) + 0.52u(k-4). &(6.3.27)\end{aligned}$$

and

$$F(q^{-1}) = 1 + 0.5q^{-1} + 0.65q^{-2}. \tag{6.3.28}$$

Then $\phi_{DC}(C_O) = 0.15$ and $\phi_{DR}(C_R) = 0.172$. Thus, $\phi_D(C_O) = 0.322$.

Figures 6.1 - 6.4 show the behaviour of the system when the sup controller is used. It can be seen that the sup controller limits the error of the system exactly to the expected bounds $\pm\phi_D(C_O)$.

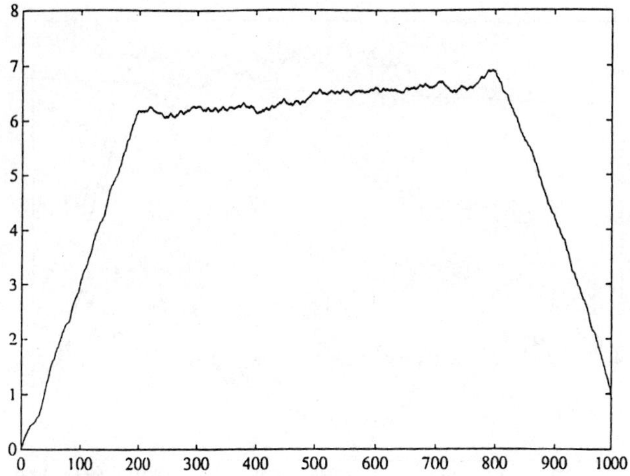

Figure 6.1: Reference input $r(k)$ of the system

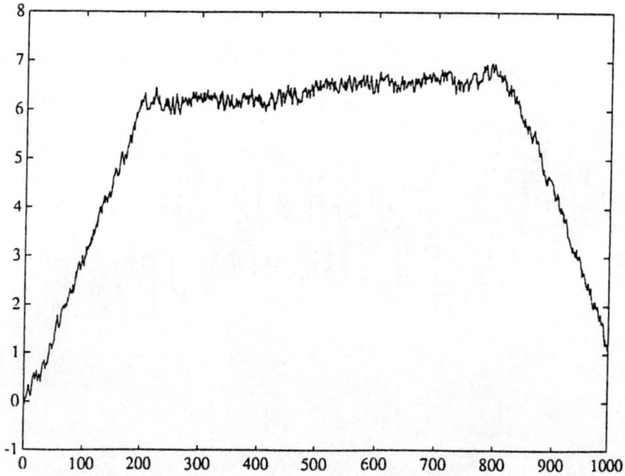

Figure 6.2: Output $y(k)$ of the system

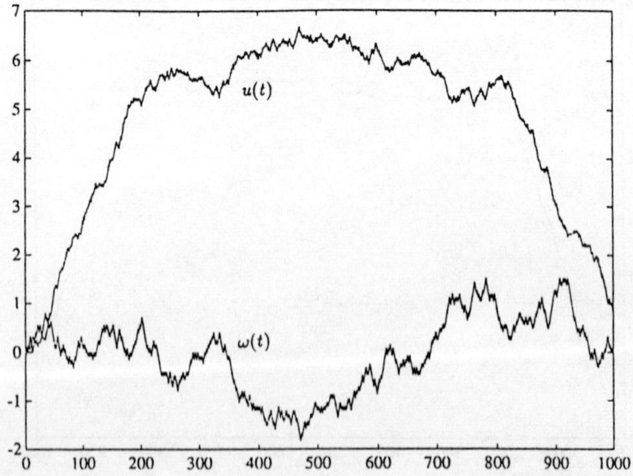

Figure 6.3: Disturbance $\omega(k)$ and control $u(k)$ of the system

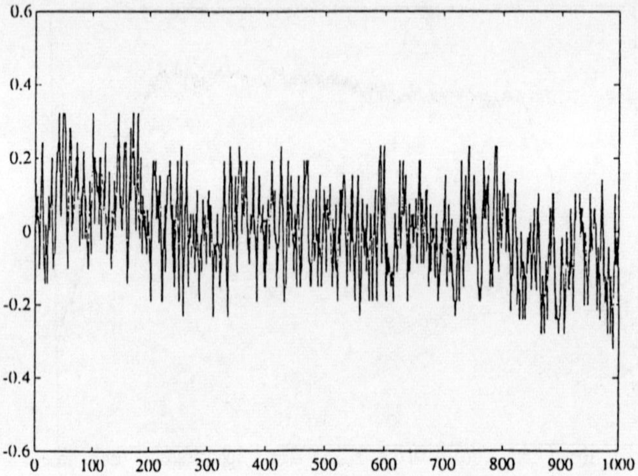

Figure 6.4: Error $e(k)$ of the system

6.4 Stability of the Closed-Loop Systems

In this section, we discuss the stability of the closed-loop systems with the sup controllers.

First, consider the regulation problem. Since the reference input $r = 0$, the plant (6.2.2) and the sup regulator (6.2.18) lead to the following closed-loop system equation:

$$\begin{aligned} y(k) &= \frac{\Delta F B_n B_d^{-1}}{\Delta B_n F + E B_n q^{-d} B_d^{-1}} \omega(k) \\ &= \frac{\Delta B_n F}{\Delta F B_d B_n + E B_n q^{-d}} \omega(k). \end{aligned} \tag{6.4.1}$$

Using (6.2.3) gives

$$y(k) = \frac{\Delta F B_n}{B_n} \omega(k). \tag{6.4.2}$$

Obviously, the characteristic equation of the closed-loop system is

$$B_n = 0. \tag{6.4.3}$$

Now consider the general case (the control problem): the reference input $r \neq 0$ and the disturbance $\omega \neq 0$. In this case, the plant (6.3.2) and the sup controller (6.3.6) result in the following closed-loop system equation:

$$\begin{aligned} e(k) &= \frac{-\Delta F B_n B_d^{-1}}{\Delta B_n F + E B_n q^{-d} B_d^{-1}} \omega(k) - \frac{B_n q^{-d} B_d^{-1} - \Delta B_n F - E B_n q^{-d} B_d^{-1}}{\Delta B_n F + E B_n q^{-d} B_d^{-1}} r(k) \\ &= \frac{-\Delta B_n F}{\Delta F B_d B_n + E B_n q^{-d}} \omega(k) - \frac{B_n q^{-d} - \Delta F B_d B_n - E B_n q^{-d}}{\Delta F B_d B_n + E B_n q^{-d}} r(k). \end{aligned} \tag{6.4.4}$$

Using (6.2.3) gives

$$e(k) = \frac{B_n}{B_n} \left((1 - q^{-d}) r(k) - \Delta F \omega(k) \right). \tag{6.4.5}$$

Clearly, the characteristic equation of the closed-loop system is

$$B_n = 0. \tag{6.4.6}$$

If the zeros of the polynomial B_n do not lie within the closed unit disc of the complex plane, then by Theorem 2.2 the system is stable and the term B_n in (6.4.2) and (6.4.5) can be cancelled without obscuring stability, so that

$$y(k) = \Delta F \omega(k), \tag{6.4.7}$$

$$e(k) = \left(1 - q^{-d}\right) r(k) - \Delta F \omega(k). \tag{6.4.8}$$

From the stability analysis above, it is concluded that the results of Sections 6.2 and 6.3 are only suited to the system $S_D(P, C)$ with a minimum phase plant. The design of the system $S_D(P, C)$ with a non-minimum phase plant will be dealt with in Section 6.6.

6.5 Uniqueness of the Control

This section establishes the uniqueness of the control generated by the sup controllers for the system $S_D(P, C)$ with a minimum phase plant.

Theorem 6.6 *If a regulator C minimises $\phi_{DR}(C)$, then*

$$\xi(k, \omega, C) = 0, \quad \forall \omega \in \mathbb{D}(0, \delta) \quad \text{and} \quad \forall k \in \mathbb{N}, \tag{6.5.1}$$

where $\xi(k, \omega, C)$ is same as $\xi(k)$ in (6.2.5).

Proof: Assume that $\xi(k, \omega^+, C) \neq 0$ for some $\omega^+ \in \mathbb{D}(0, \delta)$. Consider the particular disturbance ω^* defined by

$$\left.\begin{array}{ll} \omega^*(\tau) = \omega^+(\tau), & \tau \leq k \\ \Delta\omega^*(k + d - i) = \delta \ \mathrm{sgn}\ (f_i \xi(k, \omega^+, C)), & i = 0, 1, \ldots, d-1 \\ \Delta\omega^*(\tau) = 0, & \tau > k + d \end{array}\right\} \tag{6.5.2}$$

Obviously, $\omega^* \in \mathbb{D}(0, \delta)$.

With Lemma 6.1, it follows that

$$|y(\tau, \omega^*, C)| = \left|\xi(k, \omega^+, C)\right| + \sum_{i=0}^{d-1} |f_i| \, \delta, \quad \text{for} \quad \tau = k + d. \tag{6.5.3}$$

Hence,

$$\phi_{DR}(C) \geq |y(k + d, \omega^*, C)| > \sum_{i=0}^{d-1} |f_i| \, \delta \tag{6.5.4}$$

which, by Theorem 6.1, implies that $\phi_{DR}(C)$ is not minimal. □□

Corollary 6.1 *If a regulator C minimises $\phi_{DR}(C)$ and the input space $\mathbb{D}$ is either $\mathbb{D}(m, \delta)$ or $\mathbb{D}(N, m_0, \delta_0)$, then*

$$\xi(k, \omega, C) = 0, \quad \forall \omega \in \mathbb{D} \quad \text{and} \quad \forall k \in \mathbb{N}\,. \tag{6.5.5}$$

Proof: Immediate from the extension of Theorem 6.6. □□

Similarly, for the general case, we have the following theorem.

Theorem 6.7 *Suppose that a controller C minimises $\phi_{DR}(C)$ for the input space $\mathbb{D}_1 \times \mathbb{D}_2$, where $\mathbb{D}_1$ and $\mathbb{D}_2$ are as same as those in Section 6.3. Then*

$$\xi(k, \bar{\omega}, C) = 0, \quad \forall \bar{\omega} \in \mathbb{D}_1 \times \mathbb{D}_2 \quad \text{and} \quad \forall k \in \mathbb{N}\,. \tag{6.5.6}$$

Proof: The proof is similar to the proof for Theorem 6.6. □□

Now, first, let us establish the uniqueness of the control generated by sup regulators

Theorem 6.8 *Every regulator C that minimises $\phi_{DR}(C)$ has the same input-output map $y \mapsto u$.*

Proof: Consider the output $u^+(k)$ of a regulator C^+ that minimises $\phi_{DR}(C)$ and the output $u_R(k)$ of the sup regulator C_R. By Theorem 6.6, it follows that

$$\xi(k, \omega, C^+) = 0, \tag{6.5.7}$$

$$\xi(k, \omega, C_R) = 0. \tag{6.5.8}$$

Subtracting (6.5.8) from (6.5.7) gives

$$E\left(y^+(k) - y_R(k)\right) + \Delta F B_n \left(u^+(k) - u_R(k)\right) = 0. \tag{6.5.9}$$

where y^+ and y_R are the outputs of the plant controlled by the regulators C^+ and C_R, respectively. From the plant (6.2.2), we have,

$$B_d y^+(k) = q^{-d} B_n u^+(k) + \omega(k), \tag{6.5.10}$$

$$B_d y_R(k) = q^{-d} B_n u_R(k) + \omega(k). \tag{6.5.11}$$

Subtracting (6.5.11) from (6.5.10) gives

$$B_d\left(y^+(k) - y_R(k)\right) = q^{-d} B_n \left(u^+(k) - u_R(k)\right). \tag{6.5.12}$$

Multiplying both sides of (6.5.9) by q^{-d} gives

$$E q^{-d}\left(y^+(k) - y_R(k)\right) + \Delta F B_n q^{-d} \left(u^+(k) - u_R(k)\right) = 0. \tag{6.5.13}$$

Substitute (6.5.12) into (6.5.13) and use (6.2.3), then

$$y^+(k) = y_R(k). \tag{6.5.14}$$

From (6.5.7), (6.5.8) and (6.2.5), it follows immediately that

$$u^+(k) = -G y^+(k), \tag{6.5.15}$$

$$u_R(k) = -G y_R(k). \tag{6.5.16}$$

where $G = E(\Delta B_n F)^{-1}$.

With (6.5.14), the theorem follows. □□

The generation of this result is the following theorem for the control problem.

Theorem 6.9 *Every controller C that minimises $\phi_D(C)$ has the same input-output map $(r, y) \mapsto u$.*

Proof: Essentially as for Theorem 6.8. □□

In conclusion, Theorems 6.8 and 6.9 show that sup regulators and sup controllers generate the same control as the sup regulator C_R and the sup controller C_O for the system $S_D(P, C)$ with a minimum phase plant.

6.6 Sup Controllers for Non-Minimum Phase Plants

In this section, we deal with the design of sup controllers for the system $S_D(P, C)$ with a non-minimum phase plant and with input space $\mathbb{D}_1 \times \mathbb{D}_2$, where $\mathbb{D}_1$ and $\mathbb{D}_2$ are as same as those in Section 6.3. The performance function is

$$\phi_D(C) = \sup\left\{|e(k, \bar{w}, C)| : k \in \mathbb{N}, \bar{w} \in \mathbb{D}_1 \times \mathbb{D}_2\right\}, \tag{6.6.1}$$

where the external input vector $\bar{w} = [r, \omega]$; the error $e = r - y$; y, r and ω are the output, the reference input and the disturbance of the system, respectively.

As before, consider a plant having the ARMAX model

$$B_d y(k) = q^{-d} B_n u(k) + \omega(k), \tag{6.6.2}$$

where $B_d \in \mathbb{R}[q, \bar{n}]$ and $B_n \in \mathbb{R}[q, \bar{m}]$.

The following assumptions about the plant are made:

i) the polynomials ΔB_d and $q^{-d} B_n$ are coprime,

ii) B_n may have zeros within the closed unit disc.

By part (i) of the assumption and Proposition 2.4, there exists the following identity:

$$\Delta B_d Q + q^{-d} B_n P = 1, \tag{6.6.3}$$

where the polynomials $Q \in \mathbb{R}[q, d + \bar{m} - 1]$ with $q_0 = 1$ and $P \in \mathbb{R}[q, \bar{n}]$ are unique.

Employing the YJB parameterisation of all stabilising controllers [YJB76], the sup controller has the following form

$$\left(\Delta Q - \bar{R}\Delta B_n q^{-d}\right) u(k) = \left(P + \bar{R}\Delta B_d\right) e(k), \quad \bar{R} \in \bar{A}_1, \tag{6.6.4}$$

where P and Q are given by (6.6.3).

Combining (6.6.2) and (6.6.4), it can be seen that the error e is related to the reference input r and the disturbance ω by

$$e(k) = \left(Q - \bar{R}B_n q^{-d}\right) B_d \Delta r(k) - \left(Q - \bar{R}B_n q^{-d}\right) \Delta\omega(k). \tag{6.6.5}$$

Assume that B_n is factorised as a product BB_s, where B has all zeros inside the closed unit disc and B_s has all zeros outside the closed unit disc. Let $R = \bar{R}B_s$, where $R \in \bar{A}_1$ still makes $\bar{R} \in \bar{A}_1$ because B_s has all stable zeros. Then $Q - \bar{R}B_n q^{-d} = Q - RBq^{-d}$. Consequently, the sup controller (6.6.4) becomes

$$\left(\Delta QB_s - \Delta RB_n q^{-d}\right) u(k) = (PB_s + \Delta RB_d)\, e(k), \quad R \in \bar{A}_1, \tag{6.6.6}$$

and the relation between e, r and ω is

$$e(k) = \left(Q - RBq^{-d}\right) B_d \Delta r(k) - \left(Q - RBq^{-d}\right) \Delta\omega(k). \tag{6.6.7}$$

From the above, it is clear that the sup controller depends on the unknown polynomial $R \in \bar{A}_1$. Thus, we replace $\phi_D(C)$ by $\phi_D(R)$. The design problem is to find $R \in \bar{A}_1$ such that $\phi_D(R)$ is minimised.

Let $H^c = (Q - RBq^{-d})B_d \in \mathbb{R}[q, N_c]$ and $H^r = (Q - RBq^{-d}) \in \mathbb{R}[q, N_r]$, where $N_c \in [0, \infty]$ and $N_r \in [0, \infty]$. For the sake of simplicity, we assume that $\mathbb{D}_1 = \mathbb{D}(m_1, \delta_1)$, $\mathbb{D}_2 = \mathbb{D}(m_2, \delta_2)$ and $\min\{N_c, N_r\} > \max\{m_1, m_2\}$.

For $k \in \mathbb{N}$, let

$$|\Delta r(k-i)| = a_i\delta_1, \quad i = 0, 1, ..., N_c, \tag{6.6.8}$$

$$|\Delta\omega(k-j)| = a_j\delta_2, \quad j = 0, 1, ..., N_r. \tag{6.6.9}$$

Since $r \in \mathbb{D}(m_1, \delta_1)$ and $\omega \in \mathbb{D}(m_2, \delta_2)$,

$$a_i \geq 0, \quad i = 0, 1, ..., N_c, \tag{6.6.10}$$

$$\sum_{l=i}^{i+m_1} a_l \leq 1, \quad i = 0, 1, ..., N_c - m_1, \tag{6.6.11}$$

$$b_j \geq 0, \quad j = 0, 1, ..., N_r, \tag{6.6.12}$$

$$\sum_{l=j}^{j+m_2} b_l \leq 1, \quad j = 0, 1, ..., N_r - m_2. \tag{6.6.13}$$

Using the same reasoning as Case 1 in Section 6.2, it follows that the performance function can be expressed by

$$\phi_D(R) = \max_{\mathbf{a}} \chi_c(R, \mathbf{a})\delta_1 + \max_{\mathbf{b}} \chi_r(R, \mathbf{b})\delta_2 \tag{6.6.14}$$

where

$$\chi_c(R, \mathbf{a}) = \sum_{l=0}^{N_c} |h_l^c|\, a_l, \tag{6.6.15}$$

$$\chi_c(R, \mathbf{b}) = \sum_{l=0}^{N_r} |h_l^r|\, b_l, \tag{6.6.16}$$

$$\mathbf{a} = [a_0, a_1, \ldots, a_{N_c}], \tag{6.6.17}$$

$$\mathbf{b} = [b_0, b_1, \ldots, b_{N_r}], \tag{6.6.18}$$

and h_k^c and h_k^r are the coefficients of the polynomials H^c and H^r, respectively.

Therefore, the sup controller is obtained by

$$\min_{R \in \bar{A}_1} \left\{ \max_{\mathbf{a}} \chi_c(R, \mathbf{a})\delta_1 + \max_{\mathbf{b}} \chi_r(R, \mathbf{b})\delta_2 \right\}, \tag{6.6.19}$$

subject to (6.6.10) - (6.6.13).

In general, although it is very difficult to obtain an analytical solution for the above problem, it can be solved by numerical methods. Obviously, this is a minimax problem which consists of two loops: the internal loop (the max part) and the external loop (the min part). The internal loop,

$$\max_{\mathbf{a}} \chi_c(R, \mathbf{a})\delta_1 + \max_{\mathbf{b}} \chi_r(R, \mathbf{b})\delta_2 \tag{6.6.20}$$

subject to (6.6.10) - (6.6.13), is a linear programming problem, which can be solved by numerous algorithms, such as the simplex algorithm [Coh78, Chv83]. The external loop, $\min \phi_D(R)$ subject to $R \in \bar{A}_1$, is a problem of minimising a function.

Remark 6.3 *Setting the reference input $r = 0$ gives a special case: a regulation problem. In this case, the structure of sup regulators is same as that of the sup controllers (6.6.7). But the performance function is simpler than that of (6.6.14) and becomes*

$$\max_{\mathbf{b}} \chi_r(R, \mathbf{b})\delta_2, \tag{6.6.21}$$

subject to (6.6.12) and (6.6.13).

The sup regulator can be obtained by

$$\min_{R \in \bar{A}_1} \max_{\mathbf{b}} \chi_r(R, \mathbf{b}) \tag{6.6.22}$$

which can also be solved by the numerical methods above.

Even though it has not been proved whether there exists a bound for the order of the polynomial R such that $\phi_D(R)$ is minimised, in many cases when the order of R increases to a finite number that leads to a significant reduction of $\phi_D(R)$ then little improvement of $\phi_D(R)$ results from further increase of the order of R.

We do not intend to discuss the design of sup controllers for the complex input spaces and $\min\{N_c, N_r\} \leq \max\{m_1, m_2\}$, but the above methods could be similarly applied.

Chapter 7

Self-Tuning Sup Regulators for Systems with a Bound on the Increment of their Disturbances

7.1 Introduction

In order to control systems with unknown parameters the problem of self-tuning control has received considerable attention. Following the work of Astrom and Wittenmark [AW73], many self-tuning control methods have been studied. The principal self-tuning methods are the minimum variance self-tuner [AW73], the generalised minimum-variance approach [CG75, CG79], the pole-placement algorithm [WEPZ79, WPZ79, AW80] and the generalised predictive control approach [CMT87]. In these methods, the disturbances which the systems are subjected to are assumed to have some probabilistic and/or statistical properties, e.g. white noise with zero mean and finite variance. However, in practice, disturbances seldom meet the above assumptions, e.g. disturbances bounded on the magnitude and disturbances bounded on the increment (for example, step changes of material quality at random times). To cope with systems with disturbances characterised by a bound on their values, some robust adaptive controllers have been developed, see, [KN82, Sam83, NA86] and [OLL87].

This chapter is concerned with the problem of self-tuning control for systems with constant but unknown parameters. The disturbances to which the system may be subjected are characterised as a set of disturbances bounded on their increments. In the case of unknown parameters, since the disturbance space is not modelled as a white noise with mean zero and finite variance, the general least squares identification algorithm is not suitable to estimate the parameters. A new modified least squares algorithm (MLS) is thus developed. The self-tuning sup regulator algorithm [Liu92b] is obtained on the basis of the MLS estimator and the sup regulator developed in Chapter 6.

The main result is a characterisation of the closed-loop system obtained when the self-tuning sup regulator algorithm is applied to the system. It is shown that the properties of the self-tuning sup regulator will converge to those of the sup regulator of Chapter 6 under the assumptions that the system to be controlled is stably invertible, the time delay is known, the upper bound of the order of the system is given and the upper bound on the increment of disturbances is known.

The remainder of the chapter is organised as follows. Section 7.2 develops a new modified least squares algorithm to estimate parameters of a system with a bound on the increment of its disturbances. The self-tuning sup regulator algorithm is presented in Section 7.3. The convergence of the closed-loop system is analysed in Section 7.4. Finally, simulation results demonstrate the operation of the self-tuning sup regulator in Section 7.5.

7.2 MLS Algorithm for Parameter Estimation

Consider a plant in the system $S_D(P, C)$ without reference input, described by

$$B_d y(k) = q^{-d} B_n u(k) + \omega(k), \tag{7.2.1}$$

where the polynomials $B_d \in \mathbb{R}[q, \bar{n}]$ with $b_{d0} = 1$ and $B_n \in \mathbb{R}[q, \bar{m}]$, the time delay $d \in \mathbb{N}^+ \backslash 0$, y, u and ω are the output, the control and the disturbance of the system, respectively.

Assume the disturbance $\omega \in \mathbb{D}(0, \delta)$, which means the disturbances to which the system $S_D(P, C)$ is subjected are bounded on their increments by δ.

For a plant described by (7.2.1), if the parameters of the plant are known it is thus straightforward to obtain the sup regulator (6.2.18), i.e.

$$\Delta F B_n u(k) = -E y(k), \tag{7.2.2}$$

where polynomials $F \in \mathbb{R}[q, d-1]$ and $E \in \mathbb{R}[q, \bar{n}]$ are given by (6.2.3).

If the parameters are unknown it might be a possibility to try to determine the parameters of (7.2.1) using some identification scheme and then use the control law (7.2.2) with the true parameters substituted by their estimates. In fact, this is the basic idea of self-tuning control. Based on the least squares (LS) algorithm a modified least squares (MLS) algorithm is presented in this section to estimate the parameters of the plant (7.2.1) with a bound on the increment of disturbances.

Let y_k, u_k and ω_k denote $y(k)$, $u(k)$ and $w(k)$, respectively, and

$$G = 1 - \Delta B_d, \tag{7.2.3}$$

where

$$G = g_1 q^{-1} + g_2 q^{-2} + \ldots + g_{\bar{n}+1} q^{-\bar{n}-1}. \tag{7.2.4}$$

Then the plant (7.2.1) can be expressed as

$$y_k = \theta^T \psi_{k-1} + \Delta \omega_k, \tag{7.2.5}$$

where

$$\theta = [g_1, g_2, \ldots, g_{\bar{n}+1}, b_{n0}, b_{n1}, \ldots, b_{n\bar{m}}]^T, \tag{7.2.6}$$

$$\psi_{k-1} = [y_{k-1}, y_{k-2}, \ldots, y_{k-\bar{n}-1}, \Delta u_{k-d}, \Delta u_{k-d-1}, \ldots, \Delta u_{k-d-\bar{m}}]^T, \tag{7.2.7}$$

$$\Delta u_k = u_k - u_{k-1}, \tag{7.2.8}$$

$$\Delta \omega_k = \omega_k - \omega_{k-1}. \tag{7.2.9}$$

Note that later in this chapter we use $\hat{x}$ to denote the estimate of x. The estimation problem is then to find a θ belonging to the set defined by

$$\Xi(\theta) = \left\{\theta : \left| y_k - \theta^T \psi_{k-1} \right| \leq \delta, \forall k \in \mathbb{N} \right\}. \tag{7.2.10}$$

Three different types of solution for this problem have mainly been explored. The first alternative is to formulate the estimation problem in a geometrical setting [FH82]. The second alternative is to derive the estimation algorithm from stability consideration together with the geometrical approach [LLO87]. The third alternative is obtain the estimation algorithm from modifying the exponentially weighted recursive least squares algorithm [CdWC90]. However, there exist the following limitations in these approaches:

i) Fogel and Huang's algorithm is more complicated than the other two algorithms so that more computational work is needed.

ii) In Lozano-Leal and Ortega's algorithm,

$$\hat{\theta}_k = \hat{\theta}_{k-1} + \frac{\alpha P_{k-1}\psi_{k-1}}{1 + \psi_{k-1}^T P_{k-1} \psi_{k-1}} (|e_k| - \delta_1) \operatorname{sgn}(e_k), \quad \alpha \in (0,1), \tag{7.2.11}$$

$$P_k^{-1} = \begin{cases} P_{k-1}^{-1} + \dfrac{\alpha P_{k-1}\psi_{k-1}}{\left(1 + \psi_{k-1}^T P_{k-1}\psi_{k-1}\right) e_k} (|e_k| - \delta_1) \operatorname{sgn}(e_k), & |e_k| > \delta_1 \\ P_{k-1}^{-1}, & |e_k| \leq \delta_1 \end{cases} \tag{7.2.12}$$

$$\delta_1 = \sqrt{1+\alpha}\, \delta, \tag{7.2.13}$$

$$e_k = y_k - \hat{\theta}_{k-1}^T \psi_{k-1}. \tag{7.2.14}$$

If it is assumed that $\psi_{k-1}^T P_{k-1} \psi_{k-1}$ is finite, then this algorithm shows that

$$\lim_{k \to \infty} |e_k| = \sqrt{1+\alpha}\, \delta, \quad \alpha \in (0,1). \tag{7.2.15}$$

This means that the prediction output error e_k cannot be less than or equal to the bound δ.

iii) In Canudas and Carrillo's algorithm,

$$\hat{\theta}_k = \hat{\theta}_{k-1} + \frac{\alpha_k P_{k-1}\psi_{k-1}}{\psi_{k-1}^T P_{k-1}\psi_{k-1}} (|e_k| - \delta)\ \mathrm{sgn}(e_k), \tag{7.2.16}$$

$$P_k = \lambda^{-1}\left(P_{k-1} - \frac{\alpha_k P_{k-1}\psi_{k-1}\psi_{k-1}^T P_{k-1}}{\psi_{k-1}^T P_{k-1}\psi_{k-1}}\left(1 - \frac{\delta}{|e_k|}\right)\right), \quad \lambda \in (0,1], \tag{7.2.17}$$

$$\alpha_k = \begin{cases} 1, & |e_k| > \delta \quad \text{or} \quad \psi_{k-1}^T P_{k-1}\psi_{k-1} = 0 \\ 0, & |e_k| \le \delta \end{cases} \tag{7.2.18}$$

$$e_k = y_k - \hat{\theta}_{k-1}^T \psi_{k-1}. \tag{7.2.19}$$

In this algorithm, when $|e_k| > \delta$ and $\psi_{k-1}^T P_{k-1}\psi_{k-1}$ is very small rather than zero, the parameter estimate $\hat{\theta}_k$ might not converge because

$$P_{k-1}\psi_{k-1}\left(\psi_{k-1}^T P_{k-1}\psi_{k-1}\right)^{-1} \tag{7.2.20}$$

may be quite large. On the other hand, if setting $\alpha_k = 0$ in this case, then it is clear that it cannot guarantee that $|e_k| \le \delta$ if it is assumed that $\psi_{k-1}^T P_{k-1}\psi_{k-1}$ is finite.

In order to overcome these limitations, this section gives the following new modified least squares identification algorithm for the plant (7.2.1) on the basis of the alternatives above.

Theorem 7.1 *Consider the plant (7.2.1) and the identification algorithm:*

$$\hat{\theta}_k = \hat{\theta}_{k-1} + \frac{\alpha_k P_{k-1}\psi_{k-1}}{1 + \psi_{k-1}^T P_{k-1}\psi_{k-1}} (|e_k| - \delta)\ \mathrm{sgn}(e_k), \tag{7.2.21}$$

$$P_k = P_{k-1} - \frac{\alpha_k P_{k-1}\psi_{k-1}\psi_{k-1}^T P_{k-1}}{|e_k| + (2|e_k| - \delta)\,\psi_{k-1}^T P_{k-1}\psi_{k-1}} (|e_k| - \delta), \tag{7.2.22}$$

$$e_k = y_k - \hat{\theta}_{k-1}^T \psi_{k-1}, \tag{7.2.23}$$

$$\alpha_k = \begin{cases} 1, & |e_k| > \delta \\ 0, & |e_k| \le \delta \end{cases} \tag{7.2.24}$$

Then

$$i) \qquad \lim_{k\to\infty} \frac{\alpha_k (|e_k| - \delta)^2}{1 + \psi_{k-1}^T P_{k-1}\psi_{k-1}} = 0, \tag{7.2.25}$$

$$ii) \qquad \lim_{k\to\infty} \left|\hat{\theta}_k - \hat{\theta}_{k-1}\right| = 0, \tag{7.2.26}$$

$$iii) \qquad \|\tilde{\theta}_k\|_V^2 \le \frac{\lambda_{max}\left(P_0^{-1}\right)}{\lambda_{min}\left(P_0^{-1}\right)} \|\tilde{\theta}_0\|_V^2, \tag{7.2.27}$$

where

$$\tilde{\theta}_k = \theta - \hat{\theta}_k. \tag{7.2.28}$$

and $\lambda_{max}(\cdot)$ *and* $\lambda_{min}(\cdot)$ *denote the maximum and the minimum eigenvalues of the matrix* $(\cdot)$.

Proof:

i) Consider a Lyapunov function:

$$V_k = \tilde{\theta}_k^T P_k^{-1} \tilde{\theta}_k. \tag{7.2.29}$$

Combining (7.2.5) and (7.2.21) - (7.2.23) yields

$$\begin{aligned} V_k &= V_{k-1} + \frac{\alpha_k\left(|e_k| - \delta\right)}{\left(1 + \psi_{k-1}^T P_{k-1} \psi_{k-1}\right)|e_k|} \times \\ &\quad \times \left[\Delta\omega_k^2 - \frac{\left(1 + \psi_{k-1}^T P_{k-1}\psi_{k-1}\right)|e_k|^3}{|e_k| + \left(2|e_k| - \delta\right)\psi_{k-1}^T P_{k-1}\psi_{k-1}}\right]. \end{aligned} \tag{7.2.30}$$

For $\omega \in \mathbb{D}(0, \delta)$, then

$$\begin{aligned} V_k &\le V_{k-1} + \frac{\alpha_k\left(|e_k| - \delta\right)}{\left(1 + \psi_{k-1}^T P_{k-1} \psi_{k-1}\right)|e_k|} \times \\ &\quad \times \left[\delta^2 - \frac{\left(1 + \psi_{k-1}^T P_{k-1}\psi_{k-1}\right)|e_k|^3}{|e_k| + \left(2|e_k| - \delta\right)\psi_{k-1}^T P_{k-1}\psi_{k-1}}\right] \\ &= V_{k-1} - \frac{\alpha_k\left(|e_k| - \delta\right)}{\left(1 + \psi_{k-1}^T P_{k-1} \psi_{k-1}\right)|e_k|} \times \\ &\quad \times \frac{\left(|e_k|^3 - 2|e_k|\delta^2 + \delta^3\right)\left(1 + \psi_{k-1}^T P_{k-1}\psi_{k-1}\right) + \left(|e_k| - \delta\right)\delta^2}{|e_k| + \left(2|e_k| - \delta\right)\psi_{k-1}^T P_{k-1}\psi_{k-1}} \\ &\le V_{k-1} - \frac{\alpha_k\left(|e_k| - \delta\right)\left(|e_k|^3 - 2|e_k|\delta^2 + \delta^3\right)}{\left(|e_k| + \left(2|e_k| - \delta\right)\psi_{k-1}^T P_{k-1}\psi_{k-1}\right)|e_k|} \end{aligned} \tag{7.2.31}$$

since $|e_k| \ge \delta$. In addition, for $|e_k| \ge \delta$,

$$e_k^2 - 2\delta^2 + \frac{\delta^3}{|e_k|} \ge |e_k|\left(|e_k| - \delta\right). \tag{7.2.32}$$

It is straightforward to show that

$$V_k \le V_{k-1} - \frac{\alpha_k\left(|e_k| - \delta\right)^2}{2\left(1 + \psi_{k-1}^T P_{k-1}\psi_{k-1}\right)}. \tag{7.2.33}$$

Therefore

$$\lim_{k\to\infty} \frac{\alpha_k (|e_k| - \delta)^2}{1 + \psi_{k-1}^T P_{k-1} \psi_{k-1}} = 0. \tag{7.2.34}$$

ii) From (7.2.21), then

$$\begin{aligned} \left\| \hat{\theta}_k - \hat{\theta}_{k-1} \right\|_V^2 &= \frac{\alpha_k \psi_{k-1}^T P_{k-1}^2 \psi_{k-1}}{\left(1 + \psi_{k-1}^T P_{k-1} \psi_{k-1}\right)^2} (|e_k| - \delta)^2 \\ &\leq \frac{\alpha_k \lambda_{max}(P_{k-1})}{1 + \psi_{k-1}^T P_{k-1} \psi_{k-1}} (|e_k| - \delta)^2 . \end{aligned} \tag{7.2.35}$$

It is clear from (7.2.22) that

$$\lambda_{max}(P_k) \leq \lambda_{max}(P_{k-1}) \leq \ldots \leq \lambda_{max}(P_0). \tag{7.2.36}$$

Then (7.2.35) can be written as

$$\left\| \hat{\theta}_k - \hat{\theta}_{k-1} \right\|_V^2 \leq \alpha_k \lambda_{max}(P_0) \frac{(|e_k| - \delta)^2}{1 + \psi_{k-1}^T P_{k-1} \psi_{k-1}}, \tag{7.2.37}$$

which, together with (7.2.34), proves (ii).

iii) From the matrix inversion theorem [GS84], it follows that

$$P_k^{-1} = P_{k-1}^{-1} + \frac{\alpha_k \psi_{k-1} \psi_{k-1}^T}{1 + \psi_{k-1}^T P_{k-1} \psi_{k-1}} \left(1 - \frac{\delta}{|e_k|}\right). \tag{7.2.38}$$

Then

$$\lambda_{min}(P_k^{-1}) \geq \lambda_{min}(P_{k-1}^{-1}) \geq \ldots \geq \lambda_{min}(P_0^{-1}). \tag{7.2.39}$$

Equation (7.2.33), together with the above, gives

$$V_k \leq V_0, \tag{7.2.40}$$

$$\lambda_{min}(P_0^{-1}) \left\| \tilde{\theta}_k \right\|_V^2 \leq \lambda_{max}(P_k^{-1}) \left\| \tilde{\theta}_0 \right\|_V^2. \tag{7.2.41}$$

So,

$$\left\| \tilde{\theta}_k \right\|_V^2 \leq \frac{\lambda_{max}(P_0^{-1})}{\lambda_{min}(P_0^{-1})} \left\| \tilde{\theta}_0 \right\|_V^2. \tag{7.2.42}$$

This establishes (iii). □□

The properties (i) - (iii) will later be instrumental in establishing the stability of the self-tuning sup regulator.

7.3 The Self-Tuning Sup Regulator Algorithm

In order to control a system with constant but unknown parameters and with a bound on the increment of disturbances, a self-tuning sup regulator algorithm is presented on the basis of the sup regulator in Section 6.2 and the new MLS algorithm in Section 7.2. The following assumptions will be made about the plant (7.2.1).

Assumption 7.1

i) The time delay d is known.

ii) An upper bound for the orders of the polynomials B_d and B_n in the plant (7.2.1) is known.

iii) The polynomial B_n has all zeros outside the closed unit disc.

Note that part (iii) of Assumption 7.1 is necessary to achieve closed-loop stability for the sup regulator when the parameters are known, which is given in Chapter 6. Thus this is a natural assumption in the case where the parameters are unknown. Part (ii) of Assumption 7.1 is of significance, since it allows the order of the true plant not to be overestimated. Part (i) of Assumption 7.1 is vital to ensure $b_{n0} \neq 0$. This is needed in the subsequent development. The self-tuning sup regulator algorithm is as follows:

i) **Step 1: Parameter Estimation**

At the sampling interval t determine the parameter θ of the model

$$y_k = \theta^T \psi_{k-1} + \Delta\omega_k \tag{7.3.1}$$

using the estimation algorithm in Section 7.2 based on all data available at time t, and then compute the polynomials F and E using the following identity

$$\hat{F}_k\left(1 - \hat{G}_k\right) + \hat{E}_k q^{-d} = 1, \tag{7.3.2}$$

where $\hat{F}_k$, $\hat{G}_k$ and $\hat{E}_k$ are the estimated values of the polynomials F, G and E at time t, respectively, and $\hat{f}_0(k) = 1$.

ii) **Step 2: Control**

The objective is to design a control law such that the output satisfies

$$\lim_{k\to\infty} \sup |y_k| \leq \left\| \hat{F} \right\|_{\bar{A}_1} \delta, \tag{7.3.3}$$

where

$$\hat{F} = \lim_{k\to\infty} \hat{F}_k. \tag{7.3.4}$$

The predictable output is

$$\hat{y}_{k+d} = \hat{E}_k y_k + \hat{F}_k \hat{B}_{nk} \Delta u(k). \tag{7.3.5}$$

So, at each sampling interval determine the control variable from the sup regulator C_R^k:

$$\hat{E}_k y_k + \hat{F}_k \hat{B}_{nk} \Delta u(k) = 0, \tag{7.3.6}$$

where the polynomials $\hat{E}_k$, $\hat{F}_k$ and $\hat{B}_{nk}$ are obtained in Step 1.

In addition, the computation of the polynomials $\hat{E}_k$ and $\hat{F}_k$ can be obtained in the following manner. If $n \geq d-2$, it follows from (7.3.2) that

$$\hat{f}_1(k) = \hat{g}_1(k), \tag{7.3.7}$$
$$\hat{f}_2(k) = \hat{g}_2(k) + \hat{f}_1(k)\hat{g}_1(k-1), \tag{7.3.8}$$
$$\cdots\cdots\cdots\cdots$$
$$\hat{f}_{d-1}(k) = \hat{g}_{d-1}(k) + \hat{f}_1(k)\hat{g}_{d-2}(k-1) + \ldots + \hat{f}_{d-2}(k)\hat{g}_1(k-d+2), \tag{7.3.9}$$

and

$$\hat{e}_0(k) = \hat{g}_d(k) + \hat{f}_1(k)\hat{g}_{d-1}(k-1) + \ldots + \hat{f}_{d-1}(k)\hat{g}_1(k-d+1), \tag{7.3.10}$$
$$\hat{e}_1(k) = \hat{g}_{d+1}(k) + \hat{f}_1(k)\hat{g}_d(k-1) + \ldots + \hat{f}_{d-1}(k)\hat{g}_2(k-d+1), \tag{7.3.11}$$
$$\cdots\cdots\cdots\cdots$$
$$\hat{e}_n(k) = \hat{f}_{d-1}(k)\hat{g}_{n+1}(k-d+1). \tag{7.3.12}$$

Note that given the polynomial $\hat{G}_l$ $(l = k, k-1, \ldots)$, then $\hat{f}_i(k)$ $(i = 1, 2, \ldots, j)$ can be used to obtain $\hat{f}_{j+1}(k)$ and so on. With little computational effort, then $\hat{F}_k$ and $\hat{E}_k$ can be obtained. If $n \leq d-2$, based on (7.3.2), the polynomials $\hat{F}_k$ and $\hat{E}_k$ can also be obtained in a similar manner.

Since the estimation algorithm in Section 7.2 is computed recursively, the algorithm requires only moderate computation.

7.4 Convergence of the Algorithm

For the analysis of the convergence of the self-tuning sup regulator algorithm, the following lemma is needed.

Lemma 7.1 *If the following conditions are satisfied for a sequence $\{\varepsilon(k)\}$ and a vector sequence $\{\gamma(k)\}$:*

$$i) \quad \lim_{k\to\infty} \frac{(\varepsilon(k)-\sigma)^2}{K_1 + K_2\| \gamma(k) \|_V^2} = 0, \tag{7.4.1}$$

$$ii) \quad \| \gamma(k) \|_V \leq C_1 + C_2 \max_{\tau \leq k} \{|\varepsilon(\tau)|\}, \tag{7.4.2}$$

where $\sigma, K_1, K_2, C_1, C_2 \in [0, \infty)$, then

$$\lim_{k\to\infty} \varepsilon(k) = \sigma \tag{7.4.3}$$

and

$$\{\| \gamma(k) \|_V\} \quad \text{is} \quad \text{bounded}. \tag{7.4.4}$$

Proof: If $\{\varepsilon(k)\}$ is a bounded sequence, then by (7.4.2) $\{\| \gamma(k) \|_V\}$ is also a bounded sequence. Then by (7.4.1), it follows that

$$\lim_{k \to \infty} \varepsilon(k) = \sigma. \tag{7.4.5}$$

Now assume that $\{\varepsilon(k)\}$ is unbounded. It follows that there exists a sequence $\{k_n\}$ such that

$$\lim_{k_n \to \infty} \varepsilon(k) = \infty, \tag{7.4.6}$$

and

$$|\varepsilon(k)| \leq |\varepsilon(k_n)|, \quad \text{for} \quad k \leq k_n. \tag{7.4.7}$$

Now along the sequence $\{k_n\}$

$$\begin{aligned} \left| \frac{\varepsilon(k_n) - \sigma}{\left(K_1 + K_2 \| \gamma(k_n) \|_V^2\right)^{1/2}} \right| &\geq \frac{|\varepsilon(k_n)| - \sigma}{K_1^{1/2} + K_2^{1/2} \| \gamma(k_n) \|_V} \\ &\geq \frac{|\varepsilon(k_n)| - \sigma}{K_1^{1/2} + K_2^{1/2} (C_1 + C_2 |\varepsilon(k_n)|)}. \end{aligned} \tag{7.4.8}$$

Hence

$$\lim_{k_n \to \infty} \left| \frac{\varepsilon(k_n) - \sigma}{\left(K_1 + K_2 \| \gamma(k_n) \|_V^2\right)^{1/2}} \right| \geq \frac{1}{K_2^{1/2} C_2} > 0, \tag{7.4.9}$$

but this contradicts (7.4.1) and hence the assumption that $\{\varepsilon(k)\}$ is unbounded is false and the results follow. □□

Note that the results of Lemma 7.1 are a simplification of Lemma 6.2.1 in [GS84].

The following global convergence result of the self-tuning sup regulator algorithm can now be established.

Theorem 7.2 *Subject to Assumption 7.1, the self-tuning sup regulator algorithm (7.2.21) - (7.2.24) and (7.3.6) when applied to the plant (7.2.1) yields*

i) $y(k)$ and $u(k)$ are bounded sequences. (7.4.10)

ii) $$\limsup_{k \to \infty} |e_k| \leq \delta. \tag{7.4.11}$$

iii) $$\limsup_{k \to \infty} |y_k| \leq \left\| \hat{F} \right\|_{\bar{A}_1} \delta. \tag{7.4.12}$$

Proof: It is known from (7.2.23) that

$$e_k = y_k - \hat{\theta}_{k-1}^T \psi_{k-1}. \tag{7.4.13}$$

Let

$$\bar{y}_k = \hat{\theta}_{k-d}^T \psi_{k-1}. \tag{7.4.14}$$

Now, define a track error ε_k as

$$\varepsilon_k = y_k - \bar{y}_k \tag{7.4.15}$$

Then, from (7.4.13) and (7.4.14),

$$\varepsilon_k = \theta^T \psi_{k-1} - \hat{\theta}_{k-d}^T \psi_{k-1} = \tilde{\theta}_{k-d}^T \psi_{k-1}. \tag{7.4.16}$$

Multiplying (7.4.14) by $\hat{F}_{k-d}$ gives

$$\begin{aligned}
\hat{F}_{k-d}\bar{y}_k &= \hat{F}_{k-d}\hat{\theta}_{k-d}^T \psi_{k-1} \\
&= \hat{F}_{k-d}\hat{G}_{k-d} y_k + \hat{F}_{k-d} q^{-d}\left(\hat{B}_{nk}\Delta u(k)\right) \\
&= \hat{F}_{k-d}\left(\hat{G}_{k-d} - 1\right) y_k + \hat{F}_{k-d} y_k + \hat{F}_{k-d} q^{-d}\left(\hat{B}_{nk}\Delta u(k)\right) \\
&= \left(\hat{F}_{k-d} - 1\right) y_k + \hat{E}_{k-d} y_{k-d} + \hat{F}_{k-d} q^{-d}\left(\hat{B}_{nk}\Delta u(k)\right), \quad \text{using} \quad (7.3.2) \\
&= \left(\hat{F}_{k-d} - 1\right) y_k + q^{-d}\left(\hat{E}_k y_k + \hat{F}_k \hat{B}_{nk}\Delta u(k)\right) \\
&= \left(\hat{F}_{k-d} - 1\right) y_k, \quad \text{using} \quad (7.3.5).
\end{aligned} \tag{7.4.17}$$

Then

$$y_k = \hat{F}_{k-d}\left(y_k - \bar{y}_k\right) = \hat{F}_{k-d}\varepsilon_k. \tag{7.4.18}$$

Thus

$$|y_k| \le m_1 \max_{\tau \le t} |\varepsilon_\tau|, \tag{7.4.19}$$

where

$$m_1 = \max_{k \in \mathrm{N}} \left\| \hat{F} \right\|_{\bar{A}_1}. \tag{7.4.20}$$

Theorem 7.1, part (iii), shows that m_1 is finite.

Since the plant (7.2.1) is assumed to be a linear discrete time stably invertible system [GS84],

$$|u(k-d)| \le m_2 + m_3 \max_{\tau \le k} |y(\tau)|, \quad 0 \le m_2, m_3 < \infty. \tag{7.4.21}$$

Thus

$$\| \psi_{k-1} \|_V \le (\bar{m}+1)m_2 + (\bar{n}+\bar{m}+2)\max\{1, m_3\} \max_{\tau \le k} |y(\tau)|. \tag{7.4.22}$$

Combining (7.4.19) and (7.4.22) gives

$$\begin{aligned}
\| \psi_{k-1} \|_V &\le (\bar{m}+1)m_2 + (\bar{n}+\bar{m}+2)\max\{1, m_3\}\, m_1 \max_{\tau \le k} |\varepsilon_\tau|, \\
&\le C_1 + C_2 \max_{\tau \le k} |\varepsilon_\tau|, \quad 0 \le C_1, C_2 < \infty.
\end{aligned} \tag{7.4.23}$$

From (7.4.13) - (7.4.15), then

$$\varepsilon_k = e_k - \left(\tilde{\theta}^T_{k-1} - \tilde{\theta}^T_{k-d}\right)\psi_{k-1}. \tag{7.4.24}$$

Thus

$$\frac{\alpha_k\left|(|\varepsilon_k| - \delta)\right|}{\left(1 + \psi^T_{k-1}P_{k-1}\psi_{k-1}\right)^{1/2}} \le \frac{\alpha_k\left|(|e_k| - \delta)\right| + \alpha_k\left|\left(\tilde{\theta}^T_{k-1} - \tilde{\theta}^T_{k-d}\right)\psi_{k-1}\right|}{\left(1 + \psi^T_{k-1}P_{k-1}\psi_{k-1}\right)^{1/2}}. \tag{7.4.25}$$

Now the limit of the right-hand side is clearly zero from Theorem 7.1, parts (i) and (ii). Hence

$$\lim_{k\to\infty} \frac{\alpha_k\left(|\varepsilon_k| - \delta\right)^2}{1 + \psi^T_{k-1}P_{k-1}\psi_{k-1}} = 0. \tag{7.4.26}$$

Using (7.2.36), we have

$$\frac{\alpha_k\left(|\varepsilon_k| - \delta\right)^2}{1 + \psi^T_{k-1}P_{k-1}\psi_{k-1}} \ge \frac{\alpha_k\left(|\varepsilon_k| - \delta\right)^2}{1 + \lambda_{max}(P_0)\| \psi_{k-1} \|^2_V}. \tag{7.4.27}$$

If $\alpha_k = 0$ as $k \to \infty$, then (7.4.10) and (7.4.11) are proved. Otherwise, $\alpha_k \neq 0$ as $k \to \infty$. Thus let

$$K_1 = 1, \quad K_2 = \lambda_{max}(P_0) \quad \text{and} \quad \gamma(k) = \psi_{k-1}. \tag{7.4.28}$$

With (7.4.1), (7.4.10) and (7.4.11) follow immediately from Lemma 7.1.

Finally, with Part (ii) of Theorem 7.1,

$$\lim_{k\to\infty} \hat{F}_{k-d} = \lim_{k\to\infty} \hat{F}_k = \hat{F}. \tag{7.4.29}$$

Since $|\varepsilon_k| \to \delta$ as $k \to \infty$, it follows from (7.4.18) that

$$\lim_{k\to\infty} \sup |y_k| \le \left\| \hat{F} \right\|_{\bar{A}_1} \delta. \tag{7.4.30}$$

Therefore, the theorem is proved. □□

Conclusions

The key conclusions are that

i) Closed-loop stability is achieved;

ii) The output of the system is asymptotically limited to the bounds $\pm\left\| \hat{F} \right\|_{\bar{A}_1} \delta$.

7.5 Examples

The performance of the self-tuning sup regulator scheme is demonstrated by means of two simulation examples.

Example 7.1 ($d = 1$): Let the plant be

$$(1 + 1.9q^{-1} + 0.92q^{-2})y(k) = q^{-1}(1 + 0.85q^{-1})u(k) + \omega(k), \tag{7.5.1}$$

where the disturbance is chosen as

$$\omega(k) = \omega(k-1) + 0.1 \text{ sgn } (\nu(k)) \tag{7.5.2}$$

and $\nu(k)$ is a random variable with $E\{\nu(k)\} = 0$ and $E\{\nu^2(k)\} = 1$. From (7.5.2), clearly, the disturbance $\omega \in \mathbb{D}(0, \delta)$ with $\delta = 0.1$.

By Theorem 6.1, the sup regulator C_R for the system is

$$u(k) = 0.9y(k) - 0.98y(k-1) - 0.92y(k-2) + 0.15u(k-1) + 0.85u(k-2) \tag{7.5.3}$$

and the optimal output performance

$$\phi_{DR}(C_R) = 0.1. \tag{7.5.4}$$

From the recurrence relationship (7.3.2), it follows that

$$\hat{F}_k = 1, \quad \hat{E}_k = \left(1 - \hat{G}_k\right) q. \tag{7.5.5}$$

The regulator with the structure (7.5.3) is obtained using the self-tuning algorithm in Section 7.3.

By (7.4.12), (7.5.2) and (7.5.5),

$$\lim_{k\to\infty} \sup \phi_{DR}(C_R^k) \leq 0.1. \tag{7.5.6}$$

Figure 7.1 shows the output $y(k)$ and Figure 7.2 shows the disturbance $\omega(k)$. The control $u(k)$ is shown in Figure 7.3 and the parameter error norm $\left\| \theta - \hat{\theta}_k \right\|_V^2$ in Figure 7.4. The convergence of the parameters g_i ($i = 1, 2, 3$) and b_{ni} ($i = 0, 1$) is shown in Figure 7.5.

Example 7.2 ($d = 3$): Consider a plant

$$y(k) + 0.5y(k-1) + 0.1y(k-2) = u(k-3) + 0.8u(k-4) + \omega(k), \tag{7.5.7}$$

where the disturbance is chosen as

$$\omega(k) = \omega(k-1) + 0.08 \text{ sgn } (\nu(k)) \tag{7.5.8}$$

and $\nu(k)$ is a random variable as in Example 7.1. Clearly, the disturbance $\omega \in \mathbb{D}(0, 0.08)$.

The sup regulator C_R for the system is

$$\begin{aligned} u(k) &= -0.625y(k) - 0.31y(k-1) - 0.065y(k-2) - 0.3u(k-1) \\ &\quad +0.25u(k-2) + 0.53u(k-3) + 0.52u(k-4). \end{aligned} \tag{7.5.9}$$

By (6.2.3), the polynomial $F = 1 + 0.5q^{-1} + 0.65q^{-2}$. Hence, the optimal output performance

$$\phi_{DR}(C_R) = 0.172. \tag{7.5.10}$$

The self-tuning sup regulator with the structure (7.5.9) is obtained by the self-tuning sup regulator algorithm in Section 7.3.

When the parameters converged, $\hat{F} = 1 + 0.5004q^{-1} + 0.6503q^{-2}$. So,

$$\lim_{k\to\infty} \sup \phi_{DR}(C_R^k) \leq 0.172056. \tag{7.5.11}$$

The simulation results are shown in Figures 7.6 - 7.10.

As can be seen from the simulation results in both examples above, the outputs of the systems are limited to the bounds ± 0.1 and ± 0.172056, respectively, when the parameters converge.

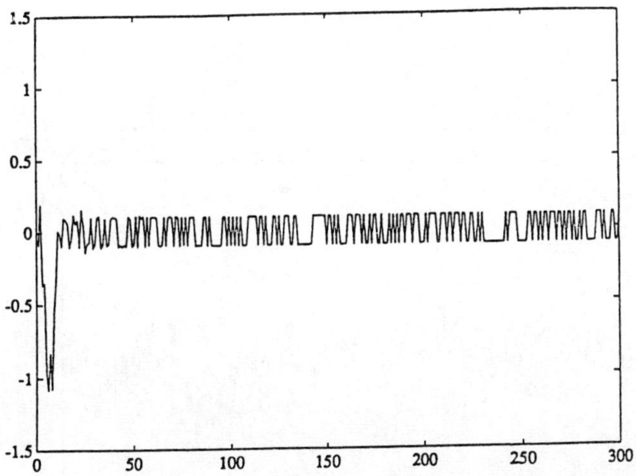

Figure 7.1: Output $y(k)$ of the system

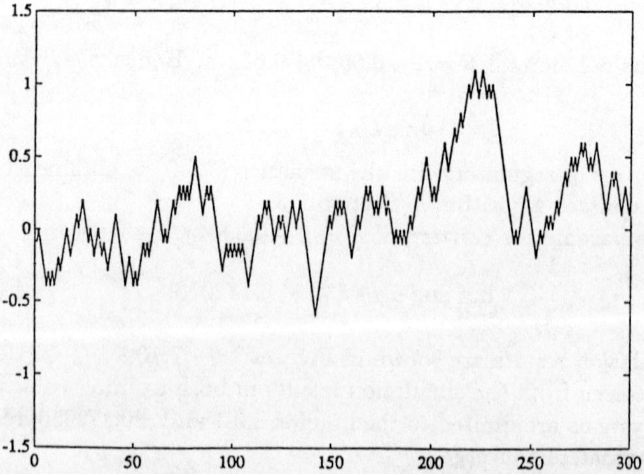

Figure 7.2: Disturbance $\omega(k)$ of the system

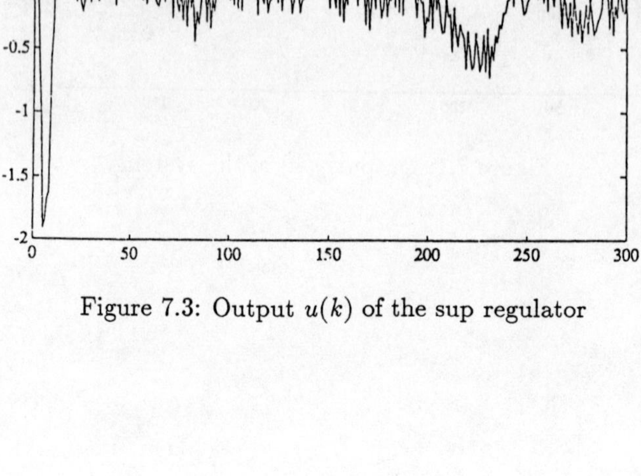

Figure 7.3: Output $u(k)$ of the sup regulator

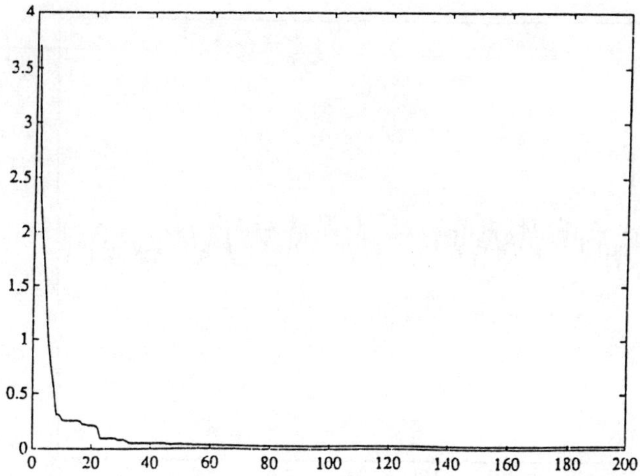

Figure 7.4: Parameter error norm $\left\| \theta - \hat{\theta}_k \right\|_2^2$

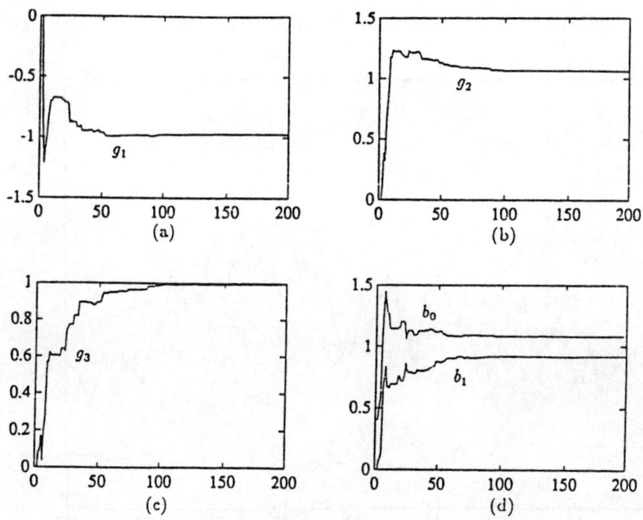

Figure 7.5: Parameters g_i $(i = 1, 2, 3)$ and b_j $(j = 1, 2)$

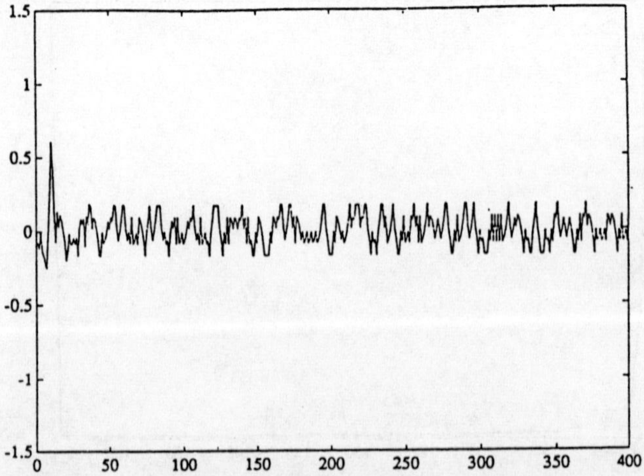

Figure 7.6: Output $y(k)$ of the system

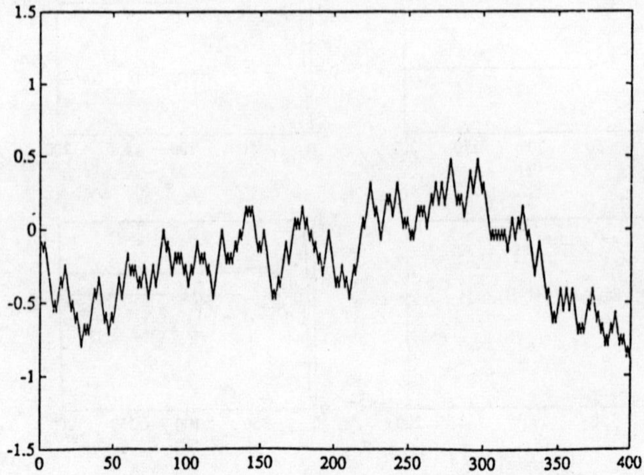

Figure 7.7: Disturbance $\omega(k)$ of the system

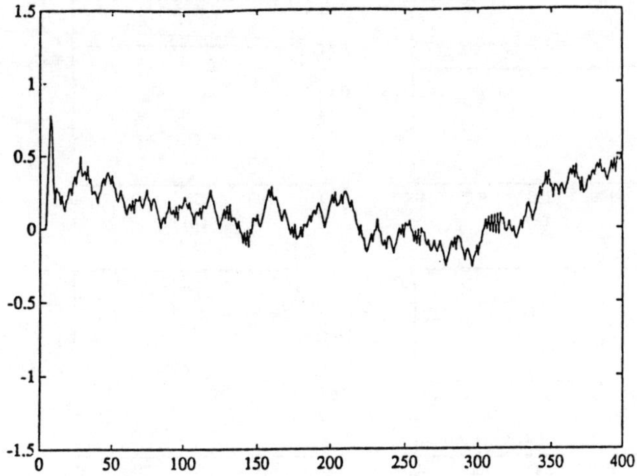

Figure 7.8: Output $u(k)$ of the sup regulator

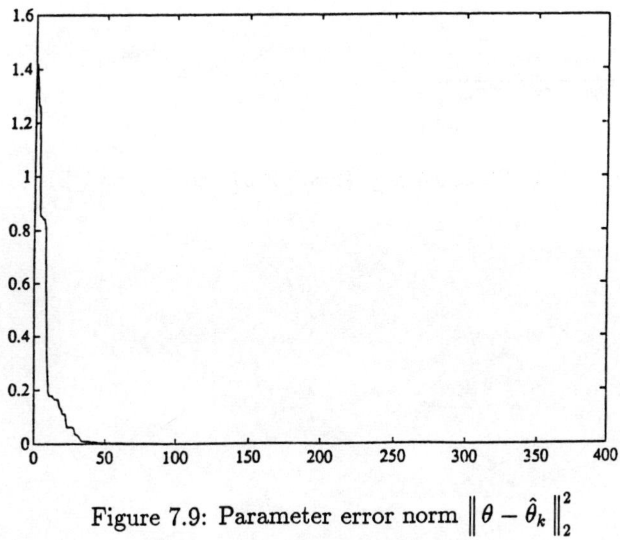

Figure 7.9: Parameter error norm $\left\| \theta - \hat{\theta}_k \right\|_2^2$

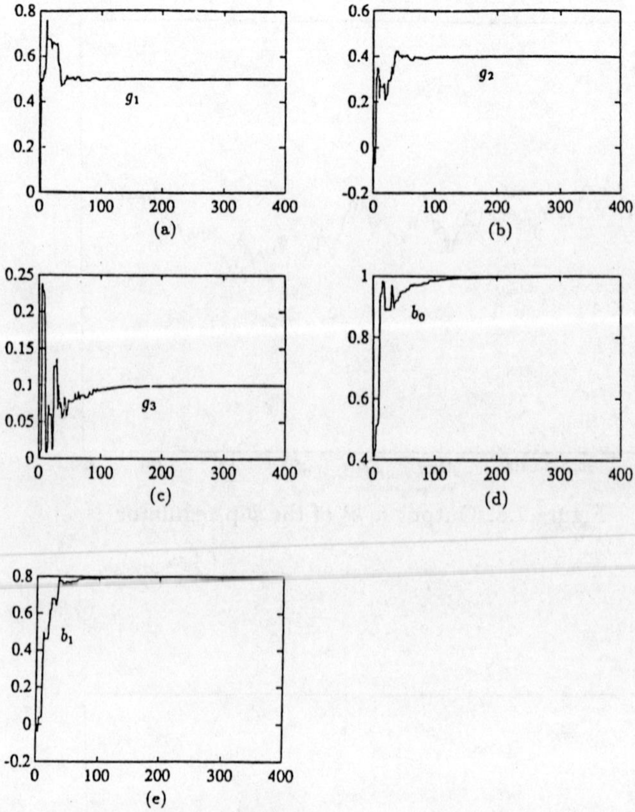

Figure 7.10: Parameters g_i $(i = 1, 2, 3)$ and b_j $(j = 1, 2)$

Part III

Continuous Time Systems

Chapter 8

Continuous Time Systems: Input Spaces and Performance Criterion

8.1 Introduction

This chapter describes the important theory behind Zakian's framework which can be used to design and analyse critical control systems in the continuous time domain. Zakian's framework is an integrated design framework which provides for the design of control systems in a methodical way. The salient elements of the framework are

- An explicit definition of performance and design criteria
- Explicit consideration of the system inputs
- Formulation of the design problem as the method of inequalities
- Solution of the design problem by numerical methods.

The inputs and disturbances are defined in terms of input spaces, defined by some bound, for example, on the supremum of the rate of change of the input. The concept of input spaces was discussed in Section 1.4. In Section 8.2, a number of useful input spaces are defined. The method of determining the supremum of the output subject to inputs belonging to this space is also derived, as are the conditions for stability.

The performance index used in Zakian's framework is defined in Section 8.3 and the design problem stated. Important concepts of stability and controllability derive from this performance index. These are defined in Section 8.4.

A number of theorems relating the input spaces to the performance are given in Section 8.5. Finally, the use of Zakian's framework in design for a standard feedback configuration is discussed in Section 8.6.

Although the preferred performance index is the error supremum, other performance indices, for example the bandwidth, may be easily incorporated into the framework.

8.2 Input Spaces in Continuous Time

A number of useful input spaces are defined in this section. Some of these spaces are the continuous time equivalents to the discrete time spaces defined in Chapter 4. The spaces $\mathcal{F}(D)$, $\mathcal{F}(m, D)$, $\mathcal{F}(N, m_0, D_0)$, $\mathcal{F}(M)$ and $\mathcal{F}(I)$ are useful for analysis and design of critical systems in the time domain, whilst the spaces $\mathcal{F}(F, D)$ and $\mathcal{F}(D_1, D_2)$ are useful in the frequency domain.

8.2.1 Input Space $\mathcal{F}(D)$

An input space characterised by a bound on the supremum of the rate of change of the input is defined as follows.

Definition 8.1 *The input w is known to the extent that it belongs to the known function space $\mathcal{F}(D)$ which is defined as the class of all functions $w : \mathbb{R} \to \mathbb{R}$ such that*

$$w^{(1)}(t) = \frac{\mathrm{d}w(t)}{\mathrm{d}t} \tag{8.2.1}$$

is piecewise continuous and where

$$w(t) = 0 \quad \forall \quad t \leq 0, \tag{8.2.2}$$

and there is also a real number D where $0 < D < \infty$, such that

$$D = \sup\left\{\left|w^{(1)}(t)\right| : t \in \mathbb{R}^+, w \in \mathcal{F}(D)\right\}. \tag{8.2.3}$$

This input space has been found to be very useful in practice, particularly for characterising reference inputs and plant disturbances.

8.2.2 Input Space $\mathcal{F}(m, D)$

The input space $\mathcal{F}(D)$ can contain a great many inputs; however, there are inputs which are not contained in this space, but which are still tolerable. An example of such an input would be one where $w^{(1)}(t)$ exceeds D, as defined in (8.2.3) for a short time, but not for long enough to drive the output to a high level. In this section, the input space $\mathcal{F}(D)$ is extended to include inputs where $w^{(1)}(t)$ may be arbitrarily large, but for a finite time, and are transient (for inputs likely to occur in practice) in the sense that $w^{(1)}(t) \to 0$ as $t \to \infty$.

Definition 8.2 *The input w is known to the extent that it belongs to the known function space $\mathcal{F}(m, D)$ which is defined as the class of all functions $w : \mathbb{R} \to \mathbb{R}$ such that $w^{(1)}$ is piecewise continuous and*

$$w(t) = 0 \quad \forall \quad t \leq 0, \tag{8.2.4}$$

and there are also real numbers m, D where $1 < m \leq \infty$ and $0 < D < \infty$, such that

$$D = \sup\left\{\left\| w^{(1)} \right\|_m : t \in \mathbb{R}^+, w \in \mathcal{F}(m, D)\right\}, \tag{8.2.5}$$

where for $1 < m < \infty$,

$$\left\| w^{(1)} \right\|_m = \left\{ \int_0^\infty \left| w^{(1)}(t) \right|^m \mathrm{d}t \right\}^{1/m}, \tag{8.2.6}$$

and for $m = \infty$,

$$\left\| w^{(1)} \right\|_\infty = \sup \left\{ \left| w^{(1)}(t) \right| : t \in \mathbb{R}^+ \right\}. \tag{8.2.7}$$

The input space $\mathcal{F}(\infty, D)$ contains inputs which are only restricted by the bound D on the absolute value of their rate of change. These inputs are described as persistent. The input space $\mathcal{F}(m, D)$, $1 < m < \infty$ contains inputs which may have an arbitrarily large rate of change, but are transient.

8.2.3 Complex Input Space $\mathcal{F}(N, m_0, D_0)$

The input space $\mathcal{F}(m, D)$ is limited in that, if $m = \infty$ it can contain persistent inputs, but a restricted range of transient inputs, and if $1 < m < \infty$ it does not contain persistent inputs. However, often the input w is made up of several parts, e.g. a persistent part and a transient part. Thus, suppose the input w is the sum of the inputs w_j $(j = 1, 2, \ldots, N)$, i.e.

$$w = \sum_{j=1}^{N} w_j, \tag{8.2.8}$$

where $w_j \in \mathcal{F}(m_{0j}, D_{0j})$. Let

$$m_0 = [m_{01}, m_{02}, \ldots, m_{0N}], \tag{8.2.9}$$
$$D_0 = [D_{01}, D_{02}, \ldots, D_{0N}]. \tag{8.2.10}$$

Definition 8.3 *The complex input space is defined as*

$$\mathcal{F}(N, m_0, D_0) = \left\{ \sum_{j=1}^{N} w_j : w_j \in \mathcal{F}(m_{0j}, D_{0j}), \quad \text{for} \quad j = 1, 2, \ldots, N \right\}. \tag{8.2.11}$$

Typically [Zak89, Zak91, Rut92], $N = 2$, $w_1 \in \mathcal{F}(\infty, D_1)$ and $w_2 \in \mathcal{F}(2, D_2)$, the persistent part is w_1 and the transient part is w_2.

8.2.4 Input Space $\mathcal{F}(M)$

Inputs characterised by a bound on their magnitude have been widely considered. In this section, such an input space is defined, and a relationship to determine the output supremum is given.

Definition 8.4 *The input w is known only to the extent that it belongs to the known function space $\mathcal{F}(M)$ which is defined as the class of all functions $w : \mathbb{R} \to \mathbb{R}$ such that*

$$w(t) = 0 \quad \forall \quad t \leq 0, \tag{8.2.12}$$

and there is a finite positive number M such that

$$M = \sup\left\{|w(t)| : t \in \mathbb{R}^+, w \in \mathcal{F}(M)\right\}. \tag{8.2.13}$$

In Section 8.6.1, this input space is shown to be of limited value.

8.2.5 Input Space $\mathcal{F}(I)$

Input space $\mathcal{F}(D)$ is shown in section 8.6.1 to be unsuitable for characterising measurement noise; a suitable input space is one characterised by a bound on the integral of the signal.

Definition 8.5 *The input w is known to the extent that it belongs to the space $\mathcal{F}(I)$ which consists of the union of all Dirac delta functions and all piecewise continuous real functions where*

$$w(t) = 0 \quad \forall \quad t \leq 0, \tag{8.2.14}$$

and there is a finite positive number I such that

$$I = \sup\left\{\left|f^{(-1)}(t)\right| : t \in \mathbb{R}^+, w \in \mathcal{F}(I)\right\}, \tag{8.2.15}$$

where

$$f^{(-1)}(t) = \int_0^t w(t)\ \mathrm{d}t. \tag{8.2.16}$$

8.2.6 Multivariable Input Space

Input space $\mathcal{F}(m, D)$ is easily extended to multi-input systems [Zak87a] with input

$$w = (w_1, w_2, \ldots, w_q), \tag{8.2.17}$$

with each input w_j belonging to a corresponding input space $\mathcal{F}(m_j, D_j)$. Let

$$1 < m_j \leq \infty, \tag{8.2.18}$$

$$0 < D_j < \infty, \tag{8.2.19}$$

$$m = (m_1, m_2, \ldots, m_q), \tag{8.2.20}$$

$$D = (D_1, D_2, \ldots, D_q), \tag{8.2.21}$$

$$\mathcal{F}(m, D) = \mathcal{F}(m_1, D_1) \times \mathcal{F}(m_2, D_2) \times \ldots \times \mathcal{F}(m_q, D_q). \tag{8.2.22}$$

The other input spaces defined in this chapter can be similarly extended.

8.2.7 Frequency Domain Input Spaces

In order to design continuous time systems by the frequency domain approaches, which are studied in Chapter 11, two input spaces in the frequency domain are introduced.

Definition 8.6 *Suppose that $W_1 \in \mathbb{H}^1(D)$. Then $W = FW_1$, where $F \in \mathbb{R}(s)$, is said to be an input in the frequency domain. The set of all functions W with a given F is called an input space, and is denoted by $\mathcal{F}(F, D)$, i.e.*

$$\mathcal{F}(F, D) = \{FW_1 : W_1 \in \mathbb{H}^1(D)\}. \tag{8.2.23}$$

For example, let

$$W = \frac{s+1}{22(s+0.1)(s+0.2)^2}. \tag{8.2.24}$$

Then W can be expressed by the product of

$$F = \frac{s+1}{s+0.1}, \tag{8.2.25}$$

$$W_1 = \frac{1}{22(s+0.2)^2}. \tag{8.2.26}$$

Obviously, $F \in \mathbb{R}(s)$ and $W_1 \in \mathbb{H}^1(0.1137)$. Therefore,

$$W \in \mathcal{F}(\frac{s+1}{s+0.1}, 0.1137). \tag{8.2.27}$$

In fact, the input space $\mathcal{F}(F, D)$ is generated by passing the set of all inputs in $\mathbb{H}^1(D)$ through the filter F; this is shown in Figure 8.1.

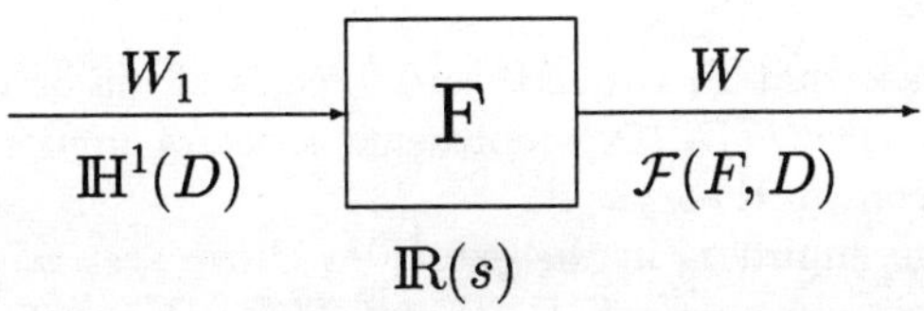

Figure 8.1: The input space $\mathcal{F}(F, D)$

Definition 8.7 *Suppose that $F_1 \in \mathbb{H}^2(D_1)$ and $F_2 \in \mathbb{H}^2(D_2)$, then $W = F_1F_2$ is said to be an input in the frequency domain. The set of all functions W is known as an input space denoted by $\mathcal{F}(D_1, D_2)$, i.e.*

$$\mathcal{F}(D_1, D_2) = \left\{F_1F_2 : F_i \in \mathbb{H}^2(D_i), i = 1, 2\right\}. \tag{8.2.28}$$

Here one can think of the input W as being generated by the given input F_2 through the transfer function F_1. Similarly, one can think of the input space $\mathcal{F}(D_1, D_2)$ as being generated by a set of given inputs F_2 with H^2-norm no larger than D_2 through a class of variable transfer functions F_1 with H^2-norm no larger than D_1. Thus, the input space $\mathcal{F}(D_1, D_2)$ can be shown as Figure 8.2.

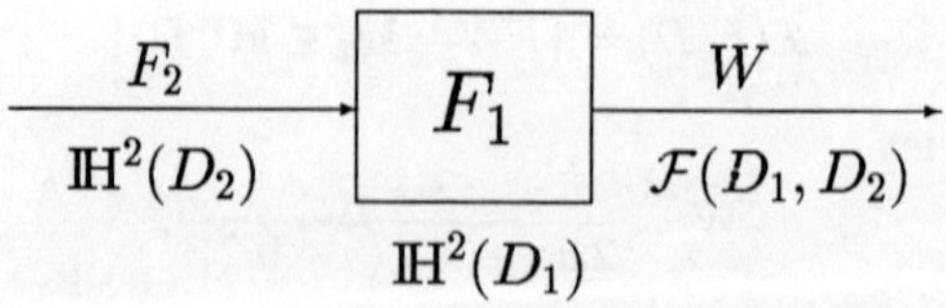

Figure 8.2: The input space $\mathcal{F}(D_1, D_2)$

From a practical viewpoint, one can also think of F_2 as a set of signals with finite power and F_1 as the transfer function of a subsystem which might be uncertain but is characterised by $\| F_1 \|_2 \leq D_1$. For example, assume that a set of given signals is $f_2 = \exp(-at)$, for $a \in [0.32, \infty)$ and $t \in \mathbb{R}^+$, and a transfer function of the subsystem with uncertainty is $F_1 = 1/(s + b)$, for $b \in [1, \infty)$. Then, obviously,

$$\| F_1 \|_2 = \frac{1}{\sqrt{2b}} \leq 0.707, \tag{8.2.29}$$

$$\| F_2 \|_2 = \frac{1}{\sqrt{2a}} \leq 1.25. \tag{8.2.30}$$

Thus, in this case, the output of the subsystem F_1 can be modelled by the space $\mathcal{F}(0.707, 1.25)$.

The special case that F_1 is in $\mathbb{H}^2$ and fixed is of major importance. In this case, the input space $\mathcal{F}(D_1, D_2)$ represents a set of inputs considered by the well-known H^∞ control theory.

Notice that the inputs w in the examples above are transient and it is conjectured that the input spaces $\mathcal{F}(F, D)$ and $\mathcal{F}(D_1, D_2)$ consist of transients.

8.2.8 Other Input Spaces

Relationships to evaluate the output supremum for several other input spaces have been determined. In Chapter 9, an input space with a bound on the second derivative of the input has been identified for one of the case studies and is defined in Appendix 9.A.

Lane [Lan92] has developed an algorithm to determine the output supremum for an input space with bounds on both the magnitude and the derivative of the input, whilst Zakian [Zak84] has developed a simple relationship which gives a majorant on the output supremum for inputs belonging to such a space. Further spaces which allow persistent inputs with a large derivative for a relatively short time have been developed by Zakian [Zak87a].

8.3 Performance Criterion

The performance measure for critical, continuous time systems is defined as

$$\phi_v(K) = \sup\left\{|v(t,w,p)| : t \in \mathbb{R}^+, w \in \mathcal{F}\right\}, \tag{8.3.1}$$

where the output $v(t,w,p)$ is the generalised output, usually the error, but it may be the control effort or other outputs, t is time, $w \in \mathcal{F}$ is the system inputs, which belong to the known function space of inputs $\mathcal{F}$, and K is the controller. ϕ_v is the greatest output as time ranges over the half line $[0,\infty)$ and the input ranges over the input space $\mathcal{F}$.

This performance index was introduced in Section 1.6.4. The reasons for its preference are :

i) it is a natural measure of performance and it has an obvious physical meaning,

ii) the system inputs are explicitly considered,

iii) it is the ideal performance index for critical systems.

From this performance index arise appropriate definitions of stability and controllability.

An appropriate mathematical statement for critical continuous time systems is expressed in the form of the following criterion:

$$\phi_v(K) \leq \varepsilon_d, \tag{8.3.2}$$

where ε_d is a finite positive number and the largest tolerable value of $\phi_v(K)$, and the design parameter is the controller K.

8.4 Stability and Controllability

Appropriate definitions of stability and controllability follow from the performance index defined by (8.3.1). These concepts of stability and controllability differ from similar traditional concepts in that the whole system description is considered; whereas the traditional concepts consider the restricted system description which excludes the environment.

Definition 8.8 *A whole system is said to be stable ϕ_v if and only if*

$$\phi_v(p) < \infty. \tag{8.4.1}$$

The concept of stability is important in Zakian's framework because numerical search methods are used to improve $\phi_v(p)$. In an unstable region Q, $\phi_v(p) = \infty \quad \forall \quad p \in Q$, so search methods cannot locate any point to improve the value of $\phi_v(p)$, and so the search cannot proceed. It is therefore necessary to stabilise the system before improvement to $\phi_v(p)$ can be made.

This concept of stability is similar to the input-output stability given by Definition 2.8; however, the concept of a system being stable ϕ_v includes consideration of the inputs. A system may be input-output stable in the sense given by Definition 2.8 but not stable ϕ_v. Additionally, the stability in the stability ϕ_v sense is dependent upon the particular characterisation of the inputs, i.e. on the environment model.

Definition 8.9 *A whole system is said to be controllable ϕ_v if and only if $\phi_v(p)$ can be made arbitrarily close to zero by appropriate choice of p.*

Controllability in the ϕ_v sense is important as it indicates whether a meaningful design problem has been formulated.

Again this concept of controllability in the ϕ_v sense is also dependent on the environment model; a system may be controllable ϕ_v for one characterisation of the inputs, but not for another.

8.5 Evaluation of Performance

In this section, relationships to determine performance are given for systems with inputs belonging to the input spaces defined in the previous section.

The Dynamical System

A system with input $w : \mathbb{R} \to \mathbb{R}$ and output $v(w) : \mathbb{R} \to \mathbb{R}$ maps an input space into an output space. The input w is known to the extent that it belongs to the known function space $\mathcal{F}$. The output is $v(w)$ and the value of the output at time t is denoted by $v(t, w)$.

8.5.1 Evaluation of Performance with Input Space $\mathcal{F}(D)$

In this section, a relationship to evaluate the performance for an input space $\mathcal{F}(D)$ is derived. Conditions for the stability of the system are derived. This input space has been found to be the most useful, and so proofs of the theorems are presented. The remaining theorems in this section have similar proofs but are not given.

It is assumed that for all $w \in \mathcal{F}(D)$ the output v is related to the input by

$$v(t, w) = \int_0^t v(t - \tau, h) w^{(1)}(\tau) \, d\tau. \tag{8.5.1}$$

The unit step response $v(h)$ is the output function where w is the unit step h. It is assumed that $v(h)$ is piecewise continuous.

Theorem 8.1 *For all $w \in \mathcal{F}(D)$, and for the output given by (8.5.1) where $v(h)$ is piecewise continuous, the performance can be calculated from*

$$\phi_v = \| v(h) \|_1 D, \tag{8.5.2}$$

where

$$\| v(h) \|_1 = \int_0^\infty |v(t,h)| \, dt. \tag{8.5.3}$$

Proof: From (8.5.1)

$$|v(t,w)| \leq \int_0^t |v(\tau,h)| \left| w^{(1)}(t-\tau) \right| \, d\tau. \tag{8.5.4}$$

From (8.2.3)

$$\sup \{ |v(t,w)| : w \in \mathcal{F}(D) \} \leq D \int_0^t |v(\tau,h)| \, d\tau, \tag{8.5.5}$$

which gives

$$\sup \left\{ |v(t,w)| : t \in \mathbb{R}^+, w \in \mathcal{F}(D) \right\} \leq D \int_0^\infty |v(\tau,h)| \, d\tau. \tag{8.5.6}$$

Now consider $w_t \in \mathcal{F}(D)$ where

$$w_t^{(1)}(t-\tau) = \begin{cases} D \text{ sgn } v(\tau,h) & \text{for} \quad 0 \leq \tau \leq t \\ 0 & \text{for} \quad \tau < 0 \end{cases} \tag{8.5.7}$$

which from (8.5.1) gives

$$v(t,w_t) = D \int_0^t |v(\tau,h)| \, d\tau. \tag{8.5.8}$$

So, as $t \to \infty$,

$$v(\infty,w_t) = D \int_0^\infty |v(\tau,h)| \, d\tau. \tag{8.5.9}$$

So from (8.5.6) and (8.5.9)

$$\phi_v = \| v(h) \|_1 D \tag{8.5.10}$$

and the theorem is proved. □□

Here $\| v(h) \|_1$ is the well-known integral absolute error (IAE) performance index.

A theorem [Zak86a] based on the definition of stability given in Section 8.4 is now derived. Let the transfer function from w to $v(w)$ be defined by

$$V(s,\delta) = sV(s,h), \tag{8.5.11}$$

where $V(s,h)$ is the Laplace transform of the step response $v(h)$ and δ is the Dirac delta function. If w is Laplace transformable, as it is if $w \in \mathcal{F}(D)$, then (8.5.1) gives $V(s,w) = V(s,\delta)W(s)$ where $W(s)$ is the transform of $w(t)$.

Theorem 8.2 *Let the $V(s,\delta)$ be a proper rational function of s and let Λ denote the set of all its finite poles λ. For all $w \in \mathcal{F}(D)$, let the output $v(t,w)$ be given by (8.5.1). Then*

$$\phi_v < \infty \tag{8.5.12}$$

if and only if the following two conditions hold [Zak86a]:

$$a) \qquad V(0,\delta) = 0, \tag{8.5.13}$$

$$b) \qquad \mathrm{Re}\{\lambda\} < 0 \quad \forall \quad \lambda \in \Lambda. \tag{8.5.14}$$

Moreover, the relation in (8.5.2) of Theorem 8.1 holds.

Proof: Since $v(h)$ is Laplace transformable, $\frac{1}{s}V(s,\delta) = V(s,h)$ and since

$$\lim_{s\to 0} V(s,\delta) = 0, \tag{8.5.15}$$

it then follows that $V(s,h)$ does not have a pole at $s = 0$, and in fact, $V(s,\delta)$ and $V(s,h)$ have the same poles. Consequently, if the poles of $V(s,\delta)$ have negative real parts then $\| v(h) \|_1 < \infty$. Therefore by (8.5.2), $\phi_v < \infty$.

Conversely, if $\phi_v < \infty$ then by (8.5.2), $\| v(h) \|_1 < \infty$ which implies that (8.5.14) holds. Also, $\| v(h) \|_1 < \infty$ implies that $v(\infty, h) = 0$ which by the final value theorem of the Laplace transform, implies that $V(0,\delta) = 0$. □□

Condition (a) is in addition to the condition stated in Proposition 2.5, which is condition (b).

8.5.2 Evaluation of Performance with Input Space $\mathcal{F}(m, D)$

In this section, a relationship to evaluate performance is given for an input space $\mathcal{F}(m,D)$.

It is assumed that for all $w \in \mathcal{F}(m,D)$ the output is related to the input by (8.5.1). Again the unit step response $v(h)$ is the output function where w is the unit step h. $v(h)$ is assumed to be piecewise continuous.

Theorem 8.3 *The performance can be calculated from*

$$\phi_v = \| v(h) \|_n D, \tag{8.5.16}$$

where

$$\| v(h) \|_n = \left\{ \int_0^\infty |v(t,h)|^n \, d\tau \right\}^{1/n}, \tag{8.5.17}$$

provided that

$$\frac{1}{m} + \frac{1}{n} = 1. \tag{8.5.18}$$

The proof can be found in [Zak86b].

Here $\| v(h) \|_2$ is the well-known quadratic index, sometimes called the root integral square error (RISE).

8.5.3 Evaluation of Performance with Complex Input Space

In this section, a relationship to evaluate the performance is given for the complex input space $\mathcal{F}(N, m_0, D_0)$.

It is assumed that for all $w_j \in \mathcal{F}(m_{0j}, D_{0j})$ the output is related to the input by

$$v(t, w_j) = \int_0^t v(t-\tau, h) w_j^{(1)}(\tau)\, d\tau. \tag{8.5.19}$$

Again, the unit step response $v(h)$ is the output function where w is the unit step h. It is assumed that $v(h)$ is piecewise continuous.

Theorem 8.4 *The performance can be calculated from*

$$\phi_v = \sum_{i=1}^{N} \| v(h) \|_{n_{0j}} D_{0j}, \tag{8.5.20}$$

where

$$\| v(h) \|_{n_{0j}} = \left\{ \int_0^\infty |v(t,h)|^{n_{0j}}\, d\tau \right\}^{1/n_{0j}}, \tag{8.5.21}$$

provided that

$$\frac{1}{m_{0j}} + \frac{1}{n_{0j}} = 1. \tag{8.5.22}$$

Proof: The proof is a particular case of Theorem 3 in [Zak87a]. □□

Typically, $N = 2$, $w_1 \in \mathcal{F}(\infty, D_{01})$ and $w_2 \in \mathcal{F}(2, D_{02})$, so

$$\phi_v = \| v(h) \|_1 D_{01} + \| v(h) \|_2 D_{02}. \tag{8.5.23}$$

As already mentioned, $\| v(h) \|_1$ is the IAE and $\| v(h) \|_2$ is the RISE. An example of the design of a system with an input of this type is given by Rutland [Rut92].

8.5.4 Evaluation of Performance with Input Space $\mathcal{F}(M)$

In this section, a relationship to evaluate the performance is given for an input space $\mathcal{F}(M)$.

It is assumed that for all $w \in \mathcal{F}(M)$ the output is related to the input by

$$v(t, w) = \int_0^t v(t-\tau, \delta) w(\tau)\, d\tau. \tag{8.5.24}$$

The unit impulse response $v(\delta)$ is the output function where w is the unit impulse δ.

Theorem 8.5 *The performance can be calculated from*

$$\phi_v = \| v(\delta) \|_1 M, \tag{8.5.25}$$

where

$$\| v(\delta) \|_1 = \int_0^\infty |v(t, \delta)| \, \mathrm{d}t. \tag{8.5.26}$$

It is a well-known result, a brief proof can be found in [Hor63]. It is shown in the next section that, if the system reference belongs to the input space $\mathcal{F}(M)$, the error is not controllable ϕ_v. Furthermore, it is shown that if the measurement disturbance belongs to $\mathcal{F}(M)$, under certain conditions, the error is also not controllable ϕ_v. Hence this input space is less useful than, say, $\mathcal{F}(D)$ or $\mathcal{F}(m, D)$.

$\| v(\delta) \|_1$ is the L^1-norm of the system, discussed in Section 1.6.3. Analytical solutions to the problem of minimising this norm have been found [DP87c]; however, the resulting controllers are generally irrational even if the plant is rational, and therefore difficult to implement. The problem of implementing such controllers means that L^1-optimisation may be most useful in determining the limit of possible performance; thus, if the minimum possible ϕ_v is greater than ε then there is no solution to the critical problem and it has to be reformulated. For further details of the use of optimisation techniques to determine limits of performance, see [BB91].

8.5.5 Evaluation of Performance with Input Space $\mathcal{F}(I)$

In this section, a relationship to evaluate the performance is given for an input space $\mathcal{F}(I)$.

It is assumed that the input-output function $w \mapsto v(w)$ is causal, linear and time-invariant and that for all $w \in \mathcal{F}(I)$ the output is related to the input by the convolution integral

$$v(t, w) = \int_0^t v(t - \tau, \delta) w(\tau) \, \mathrm{d}\tau, \tag{8.5.27}$$

where the unit impulse response $v(t, \delta)$ is piecewise continuous and has an initial condition $v(0, \delta) = 0$. After integration by parts, (8.5.27) can be rearranged to

$$v(t, w) = \int_0^t v^{(1)}(t - \tau, \delta) f^{(-1)}(\tau) \, \mathrm{d}\tau, \tag{8.5.28}$$

where $v^{(1)}(t, \delta)$ is the derivative of the unit impulse response.

Theorem 8.6 *The performance can be calculated from*

$$\phi_v = \left\| v^{(1)}(\delta) \right\|_1 I, \tag{8.5.29}$$

where

$$\left\| v^{(1)}(\delta) \right\|_1 = \int_0^\infty \left| v^{(1)}(t, \delta) \right| \, \mathrm{d}t. \tag{8.5.30}$$

The proof can be found in [Whi90].

8.5.6 Evaluation of Performance for Multivariable Systems

Relationship (8.5.16) is easily extended to cater for multivariable systems [Zak87a] with input

$$w = (w_1, w_2, \ldots, w_q), \tag{8.5.31}$$

and output

$$v(w) = (v_1(w), v_2(w), \ldots, v_p(w)), \tag{8.5.32}$$

with each input w_j belonging to a corresponding input space $\mathcal{F}(m_j, D_j)$ as described in Section 8.2.2.

Let (v_{ij}) denote the $p \times q$ matrix (called the step response matrix) such that $v_{ij} = v_i(w)$ when $w_j = h$ and $w_k = 0$ for $k \neq j$.

A bound $\phi_{v(i)}$ on each output is now stated as

$$\phi_{v(i)} = \sup\left\{|v_i(t,w)| : t \in \mathbb{R}^+, w \in \mathcal{F}(m,D)\right\}. \tag{8.5.33}$$

It is assumed that for all $w \in \mathcal{F}(m,D)$ the output is related to the input by

$$v_i(t,w) = \sum_{j=1}^{q} \int_0^t v_{ij}(t-\tau, h) w_j^{(1)}(\tau)\, \mathrm{d}\tau. \tag{8.5.34}$$

Theorem 8.7 *For all $i = 1, 2, \ldots, p$, bounds on the outputs can be calculated from*

$$\phi_{v(i)} = \sum_{j=1}^{q} \| v_{ij}(h) \|_{n_j} D_j, \tag{8.5.35}$$

where the step response $v_{ij}(h)$ is piecewise continuous for all $i = 1, 2, \ldots, p$ and $j = 1, 2, \ldots, q$, provided

$$\frac{1}{m_j} + \frac{1}{n_j} = 1 \quad \forall \quad j = 1, 2, \ldots, q. \tag{8.5.36}$$

The proof can be found in [Zak87a].

8.6 The Standard Feedback Problem

The standard feedback continuous system $S_C(G,K)$ is shown in Figure 2.2. The plant is assumed fixed and known, and represented by the strictly proper transfer function $G(s)$, with an input, also known as the control, u, and an output y. The controller has a proper transfer function $K(s)$, which is the design parameter. The system has input $\bar{w} = (\psi, \omega, r)$ consisting of measurement noise ψ, plant disturbance ω, and a reference r which is to be followed by the plant output. The system has error $e = r - y$.

Assuming the system is linear, and taking Laplace transforms of the signals, from Figure 2.2,

$$\begin{aligned} Y(s) &= \Omega(s) + G(s)K(s)\left[R(s) - \Psi(s) - Y(s)\right] \\ &= S(s)\Omega(s) + T(s)\left[R(s) - \Psi(s)\right], \end{aligned} \tag{8.6.1}$$

where the sensitivity function $S(s)$ is defined as

$$S(s) = \left[1 + G(s)K(s)\right]^{-1}, \tag{8.6.2}$$

and the closed-loop transfer function $T(s)$ is defined as

$$T(s) = S(s)G(s)K(s). \tag{8.6.3}$$

The error $E(s)$ is thus,

$$\begin{aligned} E(s) &= R(s) - S(s)\Omega(s) - T(s)\left[R(s) - \Psi(s)\right] \\ &= \left[1 - T(s)\right]R(s) - S(s)\Omega(s) + T(s)\Psi(s), \end{aligned} \tag{8.6.4}$$

but from (8.6.2) and (8.6.3)

$$T(s) + S(s) = 1, \tag{8.6.5}$$

so

$$E(s) = S(s)\left[R(s) - \Omega(s)\right] + T(s)\Psi(s). \tag{8.6.6}$$

The sensitivity function and the closed-loop transfer function are both dependent on the controller K which is the design parameter. From (8.6.6), in order to design to reduce the error, both $S(s,K)$ and $T(s,K)$ need to be made small in some sense by choice of K. However, from (8.6.5) it can be seen that if $S(s,K)$ is made nearly zero, $T(s,K)$ becomes nearly unity; and conversely, if $T(s,K)$ is nearly zero, $S(s,K)$ is nearly unity. There is thus an unavoidable trade-off between attenuating plant disturbances, and filtering out measurement error. This inevitable conflict incorporated into (8.6.5) means that $T(s)$ is often known as the complementary sensitivity function, which also distinguishes it from other closed-loop transfer functions.

There is less conflict involved between the need to reduce the effect on the error of ψ and of r because pre-filtering can often be introduced on r, and a feedback loop filter introduced.

If the measurement error can be completely eliminated, then it would appear that the design process becomes relatively easy. However, there is another potential source of conflict within the system. In most control systems, it is a requirement that the control u be kept sufficiently small. From Figure 2.2,

$$\begin{aligned} U(s) &= K(s)\left[R(s) - \Psi(s) - \Omega(s) - G(s)U(s)\right] \\ &= \left[1 + K(s)G(s)\right]^{-1} K(s)\left[R(s) - \Psi(s) - \Omega(s)\right]. \end{aligned} \tag{8.6.7}$$

So, from (8.6.2),

$$\begin{aligned} U(s) &= S(s)K(s)\left[R(s) - \Psi(s) - \Omega(s)\right], && (8.6.8) \\ &= \frac{T(s)}{G(s)}\left[R(s) - \Psi(s) - \Omega(s)\right]; && (8.6.9) \end{aligned}$$

from this we see that $T(s, K)$ must, in some sense, be kept small to keep u small, unless $G(s)$ is very large.

In practice, for designing in the frequency domain, the conflict between keeping $T(s, K)$ small and keeping $S(s, K)$ small is resolved by making one small at some frequencies, and the other small at other frequencies. The choice of the relevant frequency range where $T(s)$ or $S(s)$ is small depends upon the frequency distribution of the inputs. Usually, the spectra of r and ω are concentrated at low frequencies, whilst ψ extends over a much wider frequency range, and often includes significant high frequency noise. The solution which is normally adopted, therefore, is to make $|S(j\omega, K)|$ small at low frequencies and $|T(j\omega, K)|$ small at high frequencies.

What in effect happens is that the bandwidth of the closed-loop transfer function is set so that high frequency measurement noise is filtered out, and low frequency plant disturbance is filtered out. The design is thus a compromise, the choice of the bandwidth is determined by the size and frequency range of the various disturbances. If ω and ψ contain significant signals over the same frequency band, then the conflict cannot be resolved, and the performance of the control system suffers.

8.6.1 Disturbance Rejection Using Zakian's Framework

For design in the time domain, using Zakian's framework, a compromise is also achieved to resolve the conflict between the sensitivity function and the complementary sensitivity function. To an extent, the conflict between following reference inputs and filtering out measurement noise can be resolved by using a two-degree of freedom controller with a pre-compensator; however, the general arguments for a one-degree of freedom controller presented here still apply.

The inputs are known to the extent that they belong to a known function space. These are the reference input $r \in \mathcal{F}_r$, the plant disturbance $\omega \in \mathcal{F}_\omega$ and the measurement noise $\psi \in \mathcal{F}_\psi$. There are restrictions on the way these input spaces may be defined; these restrictions are dictated by the nature of $S(s)$ and $T(s)$, and are discussed prior to demonstrating how the conflict between $S(s)$ and $T(s)$ is resolved. We first need to define the lack of performance resulting from the various inputs.

Definition 8.10 *The lack of performance resulting from the reference r is*

$$\phi_{e_r} = \sup\left\{|e_r(t, r)| : t \in \mathbb{R}^+, r \in \mathcal{F}_r\right\}, \qquad (8.6.10)$$

where e_r is the error resulting from the input r.

Definition 8.11 *The lack of performance resulting from the plant disturbance* ω *is*

$$\phi_{e_\omega} = \sup\left\{|e_\omega(t,\omega)| : t \in \mathbb{R}^+, \omega \in \mathcal{F}_\omega\right\}, \tag{8.6.11}$$

where e_ω *is the error resulting from the input* ω.

Definition 8.12 *The lack of performance resulting from the measurement noise* ψ *is*

$$\phi_{e_\psi} = \sup\left\{|e_\psi(t,\psi)| : t \in \mathbb{R}^+, \psi \in \mathcal{F}_\psi\right\}, \tag{8.6.12}$$

where e_ψ *is the error resulting from the input* ψ.

The Reference and Plant Disturbance Input Spaces

If the input spaces $\mathcal{F}_\omega$ and $\mathcal{F}_r$ are characterised by $\mathcal{F}_\omega(M_\omega)$ and $\mathcal{F}_r(M_r)$, respectively, as defined by (8.2.13), then the performance error generated by the reference input and plant disturbance cannot be made arbitrarily close to zero, and the system $S_C(G,K)$ is thus not controllable in the controllable ϕ sense. This is stated in the following two propositions.

The sensitivity function $S(s)$ is the operator which generates the errors e_r and e_ω from the inputs r and ω respectively. The inverse Laplace transform of $S(s)$ is denoted by $e_S(\delta)$.

Proposition 8.1 *For* $\omega \in \mathcal{F}_\omega(M_\omega)$, $S_C(G,K)$ *is not controllable* ϕ_{e_ω}.

Proof: From (8.6.2), the sensitivity function is

$$S(s) = \frac{1}{1+K(s)G(s)}, \tag{8.6.13}$$

which is of the form

$$S(s) = \frac{A(s)}{B(s)}, \tag{8.6.14}$$

where

$$A(s) = s^n + a_1 s^{n-1} + \ldots + a_{n-1}s + a_n, \tag{8.6.15}$$

$$B(s) = s^n + b_1 s^{n-1} + \ldots + b_{n-1}s + b_n. \tag{8.6.16}$$

Thus $S(s)$ is proper, but not strictly proper. From (8.6.1)

$$S(s) = 1 + \frac{A(s)-B(s)}{B(s)}, \tag{8.6.17}$$

$$e_S(t,\delta) = \delta(0) + \mathcal{L}^{-1}\left[\frac{A(s)-B(s)}{B(s)}\right]. \tag{8.6.18}$$

Hence

$$\| e_S(\delta) \|_1 \geq 1, \tag{8.6.19}$$

so from (8.5.25)

$$\phi_{e_\omega} \geq M_\omega. \tag{8.6.20}$$

Hence ϕ_{e_ω} cannot be made arbitrarily close to zero, so $S_C(G,K)$ is uncontrollable ϕ_{e_ω}. □□

Proposition 8.2 *For $r \in \mathcal{F}_r(M_r)$, $S_C(G,K)$ is not controllable ϕ_{e_r}.*

Proof: As for Proposition 8.1. □□

However, if the input spaces $\mathcal{F}_r$ and $\mathcal{F}_\omega$ are characterised by $\mathcal{F}_\omega(D_\omega)$ and $\mathcal{F}_r(D_r)$, respectively, as defined by Definition 8.1, then the system is controllable in the controllable ϕ sense. It is thus practicable to use $\mathcal{F}(D)$ to characterise $\mathcal{F}_r$ and $\mathcal{F}_\omega$, provided integral action is present to satisfy the stability condition (8.5.13).

The fact that the errors are uncontrollable in the ϕ sense if the input space is of the $\mathcal{F}(M)$ type can be looked upon as the output being unable to follow an input which is changing infinitely fast. $\mathcal{F}(M)$ contains inputs which do so.

The Measurement Noise Input Space

If the input space $\mathcal{F}_\psi$ is characterised by $\mathcal{F}_\psi(D_\psi)$ as defined by Definition 8.1, then the performance ϕ_{e_ψ} generated by the measurement disturbance ψ is not finite, and so $S_C(G,K)$ is unstable ϕ.

The complementary sensitivity function $T(s)$ is the operator which generates the errors e_ψ from the input ψ. The inverse Laplace transform of $T(s)$ is denoted by $e_T(\delta)$.

Proposition 8.3 *For $\psi \in \mathcal{F}_\psi(D_\psi)$, $S_C(G,K)$ is unstable ϕ_{e_ψ} provided $K(s)G(s)$ has no open-loop zeros at $s = 0$.*

Proof: From (8.6.3)

$$T(s) = \frac{K(s)G(s)}{1 + K(s)G(s)}. \tag{8.6.21}$$

This can be put into the form

$$T(s) = \frac{A(s)}{B(s)}, \tag{8.6.22}$$

where

$$A(s) = a_n s^n + a_{n-1}s^{n-1} + \ldots + a_1 s + a_0, \qquad a_0 > 0, \tag{8.6.23}$$

$$B(s) = b_m s^m + b_{m-1}s^{m-1} + \ldots + b_1 s + 1, \tag{8.6.24}$$

and

$$m > n. \tag{8.6.25}$$

Even if $K(s)G(s)$ has a pole at $s = 0$ (i.e. integral action), $T(s)$ cannot have a zero at $s = 0$, and so stability condition (8.5.13) cannot be satisfied, and hence $S_C(G, K)$ is unstable ϕ_{e_ψ}. □□

The measurement disturbance ψ is in fact the difference between the actual output and the measured output. The input space $\mathcal{F}(D)$ contains inputs that are infinite in magnitude, thus the measured output is infinitely different to the actual output, which obviously makes the control system useless. Thus the space $\mathcal{F}(D)$ is not a good characterisation of $\mathcal{F}_\psi$.

However, if the input space $\mathcal{F}_\psi$ is characterised by $\mathcal{F}_\psi(M_\psi)$ as defined by Definition 8.4, then (8.5.13) is no longer a condition for stability, so stability is possible and (8.5.25) can be used to evaluate ϕ_{e_ψ}. However, the system may not be controllable ϕ.

Proposition 8.4 *For $\psi \in \mathcal{F}_\psi(M_\psi)$ and $K(s)G(s)$ has at least one open-loop pole at $s = 0$, $S_C(G, K)$ is uncontrollable ϕ_{e_ψ}.*

Proof: If integral action is present, then from (8.6.21), $a_0 = 1$ in (8.6.1), so, from the final value theorem,

$$e_T(\infty, h) = 1, \tag{8.6.26}$$

so from (8.5.25)

$$\phi_{e_\psi} \geq M_\psi, \tag{8.6.27}$$

and hence ϕ_{e_ψ}, although finite, cannot be made arbitrarily close to zero, so $S_C(G, K)$ is uncontrollable ϕ_{e_ω}. □□

However, if the input space $\mathcal{F}_\psi$ is characterised by $\mathcal{F}_\psi(I_\psi)$ as defined by Definition 8.5, then $S_C(G, K)$ is both stable ϕ and controllable ϕ.

8.6.2 Performance Evaluation and Design

The overall performance of the system is now defined.

Definition 8.13 *For $\bar{w} \in \mathcal{F}$, the performance is*

$$\phi_e = \sup\left\{|e(t, \bar{w})| : t \in \mathbb{R}^+, \bar{w} \in \mathcal{F}\right\}, \tag{8.6.28}$$

where $\bar{w} = (\psi, \omega, r)$ and $\mathcal{F} = \mathcal{F}_\psi \times \mathcal{F}_\omega \times \mathcal{F}_r$.

The following theorem enables the overall performance to be simply calculated.

Theorem 8.8 *For the linear system $S_C(G, K)$ and $\bar{w} \in \mathcal{F}$, the performance can be evaluated from*

$$\phi_e = \phi_{e_\omega} + \phi_{e_r} + \phi_{e_\psi}. \tag{8.6.29}$$

Proof: Because the system is linear, this follows from Theorem 3 in Zakian [Zak87a]. □□

Typically then, if $r \in \mathcal{F}_r(D_r)$, $\omega \in \mathcal{F}_\omega(D_\omega)$ and $\psi \in \mathcal{F}_\psi(M_\psi)$, the performance of a system can be determined by

$$\phi_e(K) = \| e_T(\delta, K) \|_1 M_\psi + \| e_S(h, K) \|_1 (D_r + D_\omega), \tag{8.6.30}$$

which means that the step response of the sensitivity function must be made small and the impulse response of the complementary sensitivity function must be made small, the respective sizes of M_ψ and $(D_r + D_\omega)$ determining the balance between them. The conflict between $S(s, K)$ and $T(s, K)$ is resolved to some extent by the fact that the impulse response is the derivative of the step response.

Alternatively, if a system which is controllable ϕ_e is required, or if $\psi \in \mathcal{F}(I)$ is a better characterisation for the measurement noise input space, then

$$\phi_e(K) = \left\| e_T^{(1)}(\delta, K) \right\|_1 I_\psi + \| e_S(h, K) \|_1 (D_r + D_\omega) \tag{8.6.31}$$

may be used. A discussion on the suitability of the inputs spaces $\mathcal{F}(M)$ and $\mathcal{F}(I)$ for accurately characterising measurement noise without too many fictitious inputs is included in the next section.

It should be noted that although $\mathcal{F}(I)$ provides a characterisation for the measurement noise input space which makes the system controllable ϕ_{e_ψ}, this characterisation may cause problems if there are restrictions on the control effort u. If it is required to bound ϕ_{u_ψ} where

$$\phi_{u_\psi} = \sup\left\{ |u(t, \psi)| : t \in \mathbb{R}^+, \psi \in \mathcal{F}_\psi(I) \right\}, \tag{8.6.32}$$

then the controller $K(s)$ must be strictly proper. From (8.6.8)

$$U(s) = -S(s)K(s)F_\psi(s). \tag{8.6.33}$$

The inverse Laplace transform of $S(s)K(s)$ is denoted by $u_{SK}(\delta)$. It is necessary that $S(s)K(s)$ be strictly proper so that $u_{SK}(\delta)$ is piecewise continuous. Since $S(s)$ is proper but not strictly proper, $K(s)$ must be strictly proper to ensure that $S(s)K(s)$ is strictly proper.

8.6.3 Relation Between Time and Frequency Domain Design

Design in both the frequency and the time domains requires some compromise or 'trade-off' between the sensitivity and the complementary sensitivity functions. The conflict between these is to an extent resolved in the time domain by appropriately reducing the step response of the sensitivity function, and the first or second derivative of the step response of the complementary sensitivity function, and in the frequency domain by reducing the gain over certain frequencies of the sensitivity function and reducing the gain of the complementary sensitivity function over other frequencies. The way these trade-offs are related is explained in this section.

Equation (8.6.30) can be understood in the frequency domain as follows. It is known that the bandwidth of the feedback loop is roughly inversely proportional to the speed of the step response. Thus if $(D_r + D_\omega)$ is large, this implies that a very fast step response is required to keep $\| e_S(h) \|_1$ small, and the system has a high bandwidth. The fact that $(D_r + D_\omega)$ is large means that high frequency components of $r + \omega$ are present, so a high bandwidth is required. If, on the other hand, $(D_r + D_\omega)$ is relatively small, then it becomes more important to make $\| e_T(t, \delta) \|_1$ small. Since $e_T(t, \delta)$ is the derivative of $e_T(t, h)$, then we require $\frac{\mathrm{d}}{\mathrm{d}t}(e_T(t, h))$ to be small; consequently we need $\| e_T(t, h) \|_1$ to have a comparatively sluggish response. This means that the system has a low bandwidth, which is what is required because $(D_r + D_\omega)$ being small implies that the high frequency components of $(r + \omega)$ are not present.

No inference can be made about the frequency distribution of ψ if the input space is only characterised by $\mathcal{F}(M)$. If ψ consists of low frequency drifting or bias and of small amounts of high frequency noise, then the characterisation of the input space by just M_ψ will be very close to the actual input space, and contain few fictitious inputs. However, if ψ consists of only high frequency noise with negligible bias, an input space characterised by $\mathcal{F}(M)$ will contain many fictitious inputs, and a better characterisation would be $\mathcal{F}(I)$.

A more refined input space characterisation of the measurement noise ψ is proposed in Section 8.2.5 if ψ contains both low frequency drifting or bias and large amounts of high frequency noise.

The requirement that $K(s)$ be strictly proper in order to make ϕ_{u_ψ} finite if $\mathcal{F}_\psi$ is characterised by $\mathcal{F}(I)$ is overcome in practice by including a low pass filter $H(s)$ in the feedback loop where

$$H(s) = \frac{a}{s+a}, \tag{8.6.34}$$

where a is the bandwidth of the filter. Equation (8.6.33) then becomes

$$U(s) = -S(s)K(s)H(s)\Psi(s), \tag{8.6.35}$$

which means that $S(s)K(s)H(s)$ is strictly proper because $H(s)$ is strictly proper.

8.6.4 Examples

Two simple examples demonstrate some of the limitations of using $\mathcal{F}(M)$ to characterise inputs.

Example 8.1: The plant is taken as an integrator with unity gain, and with a proportional controller with gain k, i.e. $K(s)G(s) = k/s$. This gives

$$S(s) = \frac{s}{s+k}, \tag{8.6.36}$$

and

$$T(s) = \frac{k}{s+k}. \tag{8.6.37}$$

Let $\psi \in \mathcal{F}_\psi(I_\psi)$, $\omega \in \mathcal{F}_\omega(D_\omega)$, and let $r = 0$. The performance can thus be assessed from

$$\phi_e = \left\| e_T^{(1)}(\delta) \right\|_1 I_\psi + \| e_S(h) \|_1 D_\omega. \tag{8.6.38}$$

Now

$$\begin{aligned} e_S(t,h) &= \mathcal{L}^{-1}\left[\frac{S(s)}{s}\right], \\ &= \exp(-kt), \end{aligned} \tag{8.6.39}$$

and

$$\begin{aligned} e_T^{(1)}(t,\delta) &= \mathcal{L}^{-1}\left[sT(s)\right], \\ &= k\delta(0) - k^2\exp(-kt). \end{aligned} \tag{8.6.40}$$

Taking the norms gives $\| e_S(t,h) \|_1 = 1/k$ and $\| e_T(t,\delta) \|_1 = 2k$. Thus

$$\phi_e(k) = 2kI_\psi + \frac{1}{k}D_\omega. \tag{8.6.41}$$

Thus $\phi_e(k)$ is minimised when

$$k = \sqrt{\frac{D_\omega}{2I_\psi}}. \tag{8.6.42}$$

The transmission bandwidth of the system ω_b is k. Thus

$$\omega_b = \sqrt{\frac{D_\omega}{2I_\psi}}. \tag{8.6.43}$$

Thus it can be seen that the optimum bandwidth is higher with a larger D_ω and lower with a larger I_ψ.

If however, $\psi \in \mathcal{F}_\psi(M_\psi)$, $\omega \in \mathcal{F}_\omega(M_\omega)$, and again $r = 0$, the performance is assessed from

$$\phi_e = \| e_T(\delta) \|_1 M_\psi + \| e_S(\delta) \|_1 M_\omega. \tag{8.6.44}$$

Now

$$\begin{aligned} e_S(t,\delta) &= \mathcal{L}^{-1}\left[S(s)\right], \\ &= \delta(0) - k\exp(-kt), \end{aligned} \tag{8.6.45}$$

and

$$\begin{aligned} e_T(t,\delta) &= \mathcal{L}^{-1}[T(s)], \\ &= k\exp(-kt). \end{aligned} \tag{8.6.46}$$

Taking the norms gives $\| e_S(t,\delta) \|_1 = 2$ and $\| e_T(t,\delta) \|_1 = 1$. Thus

$$\phi_e(k) = M_\psi + 2D_\omega, \tag{8.6.47}$$

and so this system is not controllable ϕ_e.

Example 8.2: The plant is taken as a first order lag with arbitrary time constant τ and unity gain, and with a proportional controller with gain k. Thus

$$K(s)G(s) = \frac{k}{\tau s + 1}. \tag{8.6.48}$$

This gives

$$S(s) = \frac{\tau s + 1}{\tau s + 1 + k}, \tag{8.6.49}$$

and

$$T(s) = \frac{k}{\tau s + 1 + k}. \tag{8.6.50}$$

Again let $\psi \in \mathcal{F}_m(M_m)$, $\omega \in \mathcal{F}_\omega(M_\omega)$, and let $r = 0$. The performance is again

$$\phi_e = \| e_T(\delta) \|_1 M_m + \| e_S(\delta) \|_1 M_\omega. \tag{8.6.51}$$

Now

$$\begin{aligned} e_S(t,\delta) &= \mathcal{L}^{-1}[S(s)], \\ &= \delta(0) - \frac{k}{\tau}\exp\left(-\frac{1+k}{\tau}t\right), \end{aligned} \tag{8.6.52}$$

and

$$\begin{aligned} e_T(t,\delta) &= \mathcal{L}^{-1}[T(s)], \\ &= \frac{k}{\tau}\exp\left(-\frac{1+k}{\tau}t\right). \end{aligned} \tag{8.6.53}$$

Taking the norms gives

$$\| e_S(t,\delta) \|_1 = 1 + \frac{k}{1+k}, \tag{8.6.54}$$

$$\| e_T(t,\delta) \|_1 = \frac{k}{1+k}. \tag{8.6.55}$$

Thus

$$\phi_e(k) = \frac{k}{1+k}M_m + \frac{1+2k}{1+k}M_\omega. \tag{8.6.56}$$

Thus $\phi_e(k)$ is dependent only upon the system gain and not the open-loop time constant.

Chapter 9

Case Studies

9.1 Altitude-Hold Autopilot

9.1.1 Introduction

In order to avoid mid-air collisions, air-traffic controllers assign to aircraft specific altitudes at which to fly. Aircraft on an easterly path are required to fly at an odd multiple of 1000 ft, while aircraft on a westerly path are required to fly at an even multiple of 1000 ft. It is therefore very important that aircraft maintain a steady altitude so as not to stray into the flight paths of other aircraft. Air-traffic controllers expect an aircraft to maintain its allotted altitude to within 50 ft.

Modern airliners, such as the Boeing 747, are fitted with an "altitude-hold autopilot". This autopilot is equipped with a control system to maintain the aircraft at a specified altitude. The requirement by air-traffic controllers that the altitude error must not exceed 50 ft means that this is a critical control system. An altitude-hold autopilot control system for a Boeing 747 airliner was designed as a case-study by Franklin, Powell and Emami-Naeini [FPEN86] using classical and state-space techniques. In this section, the system is designed using Zakian's framework, with both plant disturbances and measurement noise considered.

9.1.2 Model

The model for the longitudinal dynamics of the aircraft are based upon Newton's laws. The forces are from aerodynamic lift and drag, the propulsion system, and gravity. The forces are considered in a coordinate frame that is fixed to and rotating with the aircraft, as shown in Figure 9.1.

The model is best represented in the state-space form

$$x^{(1)}(t) \quad = \quad Ax(t) + Bu(t), \qquad (9.1.1)$$

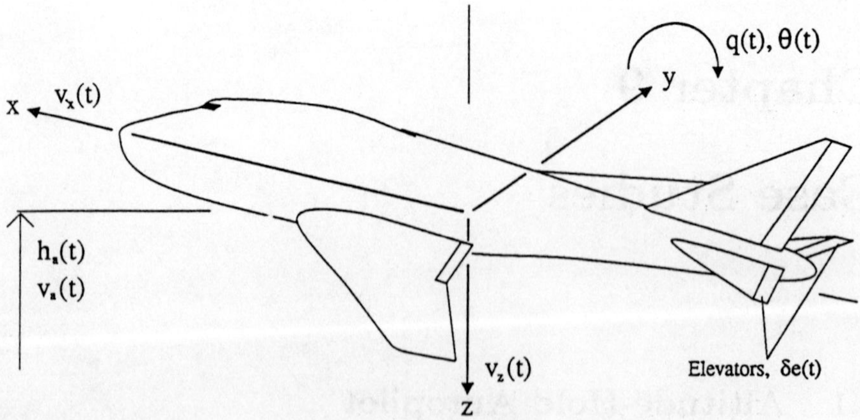

Figure 9.1: Aircraft coordinate frame

$$y(t) = Cx(t) + \omega(t), \qquad (9.1.2)$$

where $x(t)$ is a vector of the system states, A and C are matrices, $u(t)$ is the control, $y(t)$ is a vector of the measured outputs, $\omega(t)$ is a vector of the plant disturbances and B is a vector.

From [FPEN86], the linear model is

$$A = \begin{bmatrix} X_{v_x} & X_{v_z} & -V_{0_z} & -g\cos\theta_0 & 0 \\ Z_{v_x} & Z_{v_z} & V_{0_x} & -g\sin\theta_0 & 0 \\ M_{v_x} & M_{v_z} & M_q & 0 & 0 \\ 0 & 0 & 1 & 0 & 0 \\ 0 & -1 & 0 & V_{0_x} & 0 \end{bmatrix}, \qquad (9.1.3)$$

$$B = \begin{bmatrix} X_{\delta e} \\ Z_{\delta e} \\ M_{\delta e} \\ 0 \\ 0 \end{bmatrix}, \qquad (9.1.4)$$

with the state variables

$$x(t) = \begin{bmatrix} v_x(t) & v_z(t) & q(t) & \theta(t) & h_a(t) \end{bmatrix}^T, \qquad (9.1.5)$$

and the control

$$u(t) = \delta e(t), \qquad (9.1.6)$$

where:

$v_x(t)$	forward velocity perturbation in the aircraft in x direction in ft/s,
$v_z(t)$	velocity perturbation in the aircraft in z direction in ft/s,
$q(t)$	angular rate about positive y axis, or pitch rate in deg/s,
$\theta(t)$	pitch-angle perturbation with respect to the horizontal in degrees,
$h_a(t)$	aircraft altitude perturbation, the output to be controlled in feet,
$v_a(t)$	aircraft altitude rate perturbation in ft/s,
X	normalised aerodynamic force derivatives in x direction,
Z	normalised aerodynamic force derivatives in z direction,
M	normalised aerodynamic moment derivatives,
V_{0_z}	reference flight condition velocity in z axis,
V_{0_x}	reference flight condition velocity in x axis, i.e. the forward velocity of the aircraft,
g	acceleration of gravity (32.2 ft/s^2),
θ_0	angle of the x axis from horizontal in the reference flight condition,
$\delta e(t)$	movable tail section, or "elevator", angle in degrees.

The $V_{0_z}q(t)$ and $V_{0_x}q(t)$ terms are a consequence of the rotation of the coordinate frame of reference.

The aircraft will be flying horizontally when the altitude-hold autopilot is in operation, thus θ_0 is zero. For the aircraft at an altitude of 20 000 ft with a velocity V_{0_x} of 830 ft/s and with a weight of 637 000 lb, the model is [FPEN86]

$$A = \begin{bmatrix} -0.00643 & 0.0263 & 0 & -32.2 & 0 \\ -0.0941 & -0.624 & 820 & 0 & 0 \\ -0.000222 & -0.00153 & -0.668 & 0 & 0 \\ 0 & 0 & 1 & 0 & 0 \\ 0 & -1 & 0 & 830 & 0 \end{bmatrix}, \tag{9.1.7}$$

$$B = \begin{bmatrix} 0 \\ -32.7 \\ -2.08 \\ 0 \\ 0 \end{bmatrix}, \tag{9.1.8}$$

Until the early 1980's, the pitch rate $q(t)$ could be measured by means of a mechanical gyroscope. However, a more accurate measurement may now be obtained with a ring-laser gyroscope. In this device, two laser beams traverse a triangle in opposite directions. As the device rotates, the apparent frequency of the two beams shifts; this frequency is measured to obtain the rotational rate. The ring-laser gyroscope is cheaper, more reliable and more accurate than a traditional mechanical gyroscope. A similar device, known as a rate-integrating gyro, can be used to measure the pitch angle $\theta(t)$.

The altitude $h_a(t)$ may be obtained by measuring the atmospheric pressure, whilst the altitude rate $v_a(t)$ may be measured by means of a pitot tube where

$$v_a(t) = h_a^{(1)}(t) = 830\,\theta(t) - v_z(t) \text{ ft/s}. \tag{9.1.9}$$

The measurements $h_a(t)$ and $v_a(t)$ are relatively noisy, whilst the measurements $q(t)$ and $\theta(t)$ are relatively noise free.

The complete system is thus

$$\begin{aligned} x^{(1)}(t) &= Ax(t) + Bu(t), && (9.1.10)\\ y(t) &= Cx(t) + \omega(t), && (9.1.11)\\ z(t) &= y(t) + \psi(t), && (9.1.12)\\ h_a(t) &= C_h y(t), && (9.1.13) \end{aligned}$$

where $z(t)$ is a vector of the measured output signals, $\omega(t)$ is a vector of the plant disturbances and $\psi(t)$ is a vector of the measurement noise.

The outputs $y(t)$ are thus

$$y(t) = \begin{bmatrix} q(t) & \theta(t) & v_a(t) & h_a(t) \end{bmatrix}^T, \tag{9.1.14}$$

with, from (9.1.9) and (9.1.11)

$$C = \begin{bmatrix} 0 & 0 & 1 & 0 & 0 \\ 0 & 0 & 0 & 1 & 0 \\ 0 & -1 & 0 & 830 & 0 \\ 0 & 0 & 0 & 0 & 1 \end{bmatrix}. \tag{9.1.15}$$

The measured output signals, $z(t)$, are

$$z(t) = \begin{bmatrix} q_m(t) & \theta_m(t) & v_{am}(t) & h_{am}(t) \end{bmatrix}^T. \tag{9.1.16}$$

The variable to be controlled is the altitude height $h_a(t)$, so

$$C_h = \begin{bmatrix} 0 & 0 & 0 & 1 \end{bmatrix}. \tag{9.1.17}$$

The main plant disturbance will be caused by atmospheric turbulence and by up-draughts and down-draughts affecting the altitude $h_a(t)$. This disturbance is denoted by $\omega_h(t)$. The remaining disturbances are, for the purposes of this study assumed to be negligible, so the plant disturbance vector $\omega(t)$ is given by

$$\omega(t) = \begin{bmatrix} 0 & 0 & 0 & \omega_h(t) \end{bmatrix}^T. \tag{9.1.18}$$

As already stated, the measurements $h_a(t)$ and $v_a(t)$ are relatively noisy, whilst the measurements $q(t)$ and $\theta(t)$ are relatively noise free, so for this design, the measurements $q(t)$ and $\theta(t)$ are assumed to have no measurement noise. The measurement noise vector $\psi(t)$ is thus given by

$$\psi(t) = \begin{bmatrix} 0 & 0 & \psi_v(t) & \psi_h(t) \end{bmatrix}^T. \tag{9.1.19}$$

9.1.3 The Input Spaces

The plant disturbance $\omega_h(t)$ is assumed to belong to the input space $\mathcal{F}(D_h)$ as in Definition 8.1 with a corresponding bound D_h which is assumed the value

$$D_h = 33.0 \text{ ft/s.} \tag{9.1.20}$$

The measurement noise $\psi_h(t)$ is assumed to belong to the input space $\mathcal{F}(I_h)$ as in Definition 8.5 with a corresponding bound I_h which is assumed the value

$$I_h = 3.7 \text{ ft s,} \tag{9.1.21}$$

and the measurement noise $\psi_v(t)$ is assumed to belong to the input space $\mathcal{F}(I_v)$ as in Definition 8.5 with a corresponding bound I_v which is assumed the value

$$I_v = 2.0 \text{ ft.} \tag{9.1.22}$$

9.1.4 The Design

The feedback scheme is shown in Figure 9.2. Simple proportional negative feedback is used on each of the measured output signals, $z(t)$. Since the reference is constant,

$$u(t) = -K(p)z(t) \tag{9.1.23}$$

where

$$K(p) = \begin{bmatrix} p_1 & p_2 & p_3 & p_4 \end{bmatrix}. \tag{9.1.24}$$

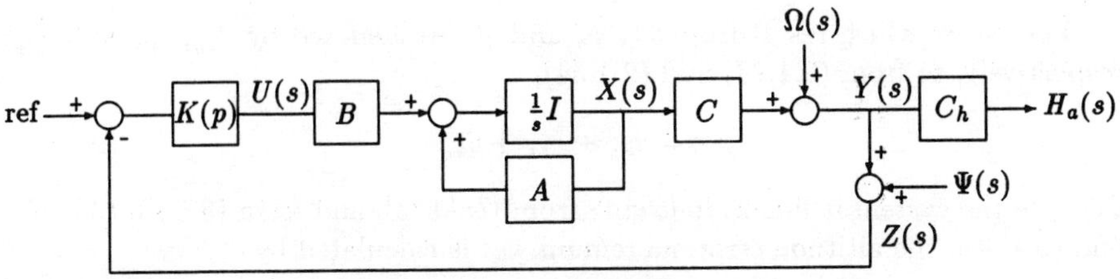

Figure 9.2: Control system configuration for altitude-hold autopilot

Taking Laplace transforms, from equations (9.1.10), (9.1.11), (9.1.12), (9.1.13) and (9.1.23), the closed-loop relationships relating the inputs to the output $h_a(t,p)$ are

$$H_a(s,p) = C_h Q(s,p)\Psi(s) \tag{9.1.25}$$

and

$$H_a(s,p) = C_h \left[I - Q(s,p)\right]\Omega(s) \tag{9.1.26}$$

where

$$Q(s,p) = -C\left[sI - A + BK(p)C\right]^{-1} BK(p), \tag{9.1.27}$$

which is put into the transfer function form

$$H_a(s,p) = S_h(s,p)\Omega_h(s) + T_v(s,p)\Psi_v(s) + T_h(s,p)\Psi_h(s), \tag{9.1.28}$$

where

$$S_h(s,p) = C_h\left[I - Q(s,p)\right]\begin{bmatrix} 0 & 0 & 0 & 1 \end{bmatrix}^T, \tag{9.1.29}$$

$$T_v(s,p) = C_h Q(s,p)\begin{bmatrix} 0 & 0 & 1 & 0 \end{bmatrix}^T, \tag{9.1.30}$$

and

$$T_h(s,p) = C_h Q(s,p)\begin{bmatrix} 0 & 0 & 0 & 1 \end{bmatrix}^T. \tag{9.1.31}$$

The design requirement is that the error does not exceed 50 feet, so the design criterion can be stated as

$$\phi_e(p) \le \varepsilon, \tag{9.1.32}$$

where

$$\varepsilon = 50 \text{ ft}. \tag{9.1.33}$$

The reference is constant, so

$$e(t) = -h_a(t). \tag{9.1.34}$$

The altitude error supremum ϕ_e is

$$\phi_e = \sup\left\{|e(t,\omega_h,\psi_h,\psi_v,p)| : t \in \mathbb{R}^+, \omega_h \in \mathcal{F}(D_h), \psi_h \in \mathcal{F}(I_h), \psi_v \in \mathcal{F}(I_v)\right\}. \tag{9.1.35}$$

The errors which result from ω_h, ψ_h and ψ_v are denoted by e_{ω_h}, e_{ψ_h} and e_{ψ_v} respectively, so from (9.1.28) and (9.1.34),

$$e = e_{\omega_h} + e_{\psi_h} + e_{\psi_v}. \tag{9.1.36}$$

Because the system is linear, following from [Zak87a], and from (8.5.2), (8.5.29) and (9.1.36), the altitude error supremum, ϕ_e, is calculated by

$$\phi_e(p) = \|\, e_{\omega_h}(h,p)\,\|_1 D_h + \left\|\, e^{(1)}_{\psi_h}(\delta,p)\,\right\|_1 I_h + \left\|\, e^{(1)}_{\psi_v}(\delta,p)\,\right\|_1 I_v, \tag{9.1.37}$$

where

$$e_{\omega_h}(t,h,p) = \mathcal{L}^{-1}\left[-\frac{1}{s}S_h(s,p)\right], \tag{9.1.38}$$

$$e^{(1)}_{\psi_h}(t,\delta,p) = \mathcal{L}^{-1}\left[-sT_h(s,p)\right], \tag{9.1.39}$$

$$e^{(1)}_{\psi_v}(t,\delta,p) = \mathcal{L}^{-1}\left[-sT_v(s,p)\right]. \tag{9.1.40}$$

The integral action between δe and h_a means that stability criterion (8.5.13) is satisfied. Expression (9.1.37) was incorporated with the moving boundaries process to find a controller $K(p)$ which satisfies the design criteria (9.1.32). The first stage was to satisfy stability criterion (8.5.14) which was achieved after a few iterations. After a further number of iterations, the design criterion (9.1.32) was satisfied with a controller

$$K = \begin{bmatrix} 30.2 & 123 & 0.672 & 1.04 \end{bmatrix}, \tag{9.1.41}$$

which gives an error supremum

$$\phi_e = 49.6 \text{ ft}, \tag{9.1.42}$$

and

$$\begin{aligned} \| e_{\omega_h}(h) \|_1 &= 0.90, & (9.1.43)\\ \left\| e^{(1)}_{\psi_h}(\delta) \right\|_1 &= 4.0, & (9.1.44)\\ \left\| e^{(1)}_{\psi_v}(\delta) \right\|_1 &= 2.58. & (9.1.45) \end{aligned}$$

The resulting closed-loop transfer functions are

$$\begin{aligned} S_h(s) &= \frac{s^5 + 42.1s^4 + 294.7s^3 + 851.5s^2 + 4.75s}{s^5 + 42.1s^4 + 260.8s^3 + 850.3s^2 + 1081.9s + 4.9}, & (9.1.46)\\ T_h(s) &= \frac{-34.0s^3 + -1.3s^2 + 1077.2s + 4.9}{s^5 + 42.1s^4 + 260.8s^3 + 850.3s^2 + 1081.9s + 4.9}, & (9.1.47)\\ T_v(s) &= \frac{-22.0s^3 + -0.84s^2 + 696.0s + 3.16}{s^5 + 42.1s^4 + 260.8s^3 + 850.3s^2 + 1081.9s + 4.9}. & (9.1.48) \end{aligned}$$

The step response $e_{dh}(t, h)$, is shown in Figure 9.3, and the closed-loop frequency response $|T_h(j\omega)|$ is shown in Figure 9.4.

A severe altitude disturbance input $f_{dh}(t) \in \mathcal{F}_{dh}(D_h)$ is shown in Figure 9.5, and severe measurement noise inputs $f_{mh}(t) \in \mathcal{F}_{mh}(I_h)$ and $f_{mv}(t) \in \mathcal{F}_{mv}(I_v)$ are shown in Figures 9.6 and 9.7 respectively. The resulting error $e(t)$ is shown in Figure 9.8, from which it can be seen that the bound ε is not exceeded.

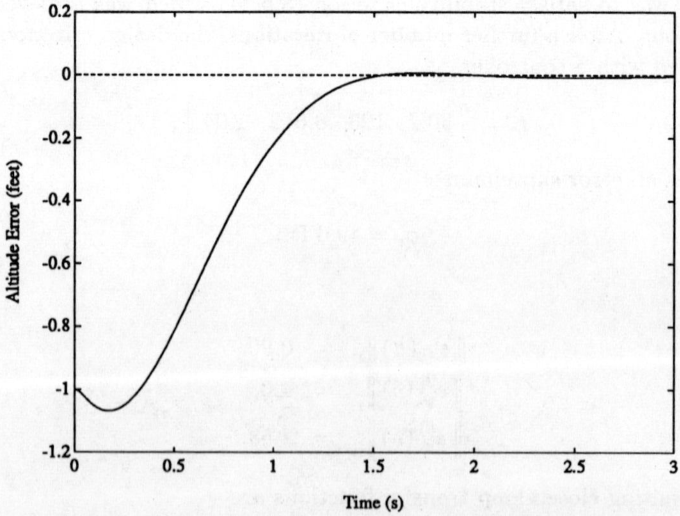

Figure 9.3: Altitude error step response $e_{\omega_h}(t,h)$

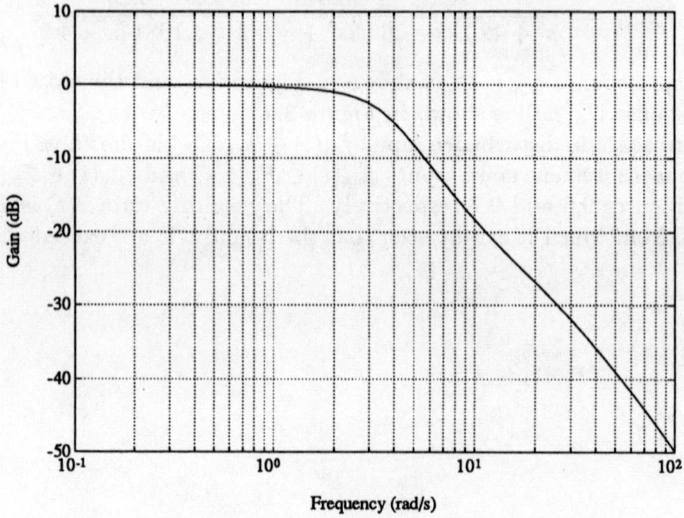

Figure 9.4: Closed loop frequency response $|T_h(j\omega)|$

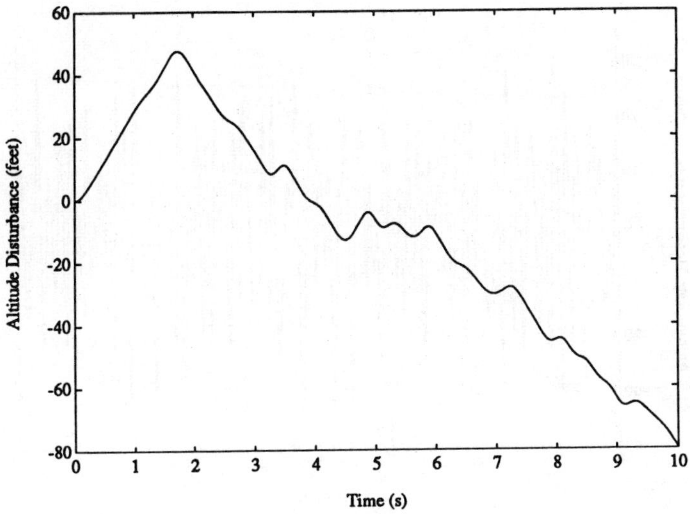

Figure 9.5: Plant disturbance input $\omega_h(t)$

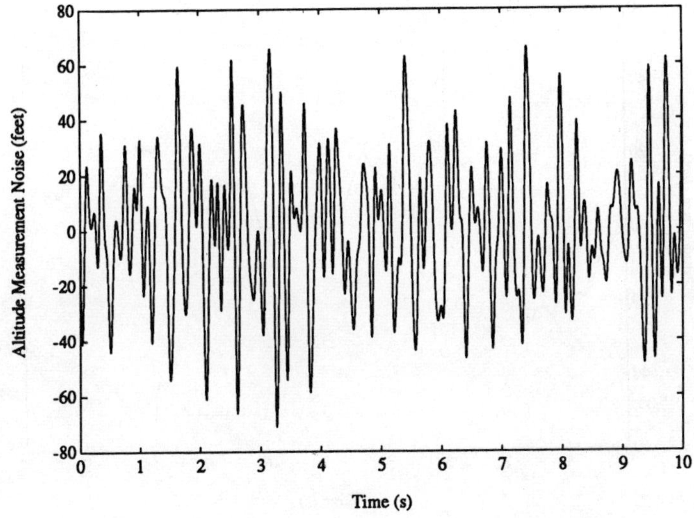

Figure 9.6: Altitude measurement noise $\psi_h(t)$

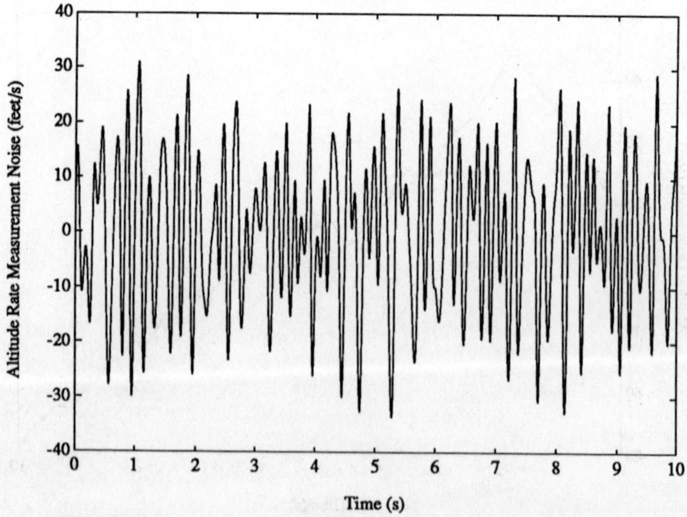

Figure 9.7: Altitude rate measurement noise $\psi_v(t)$

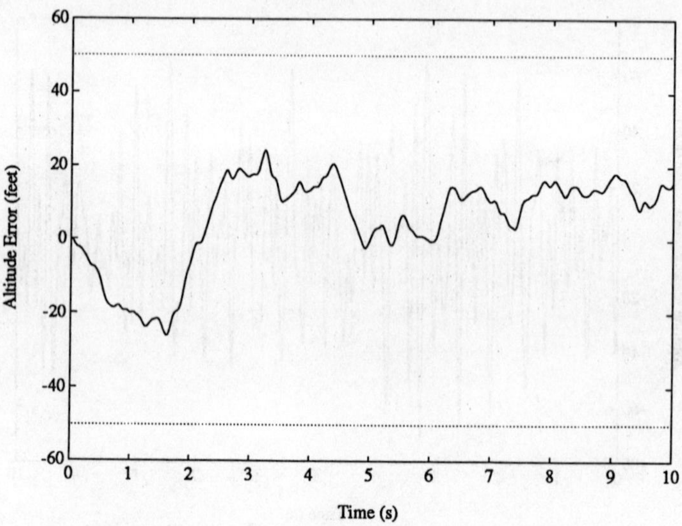

Figure 9.8: Altitude error response $e(t)$ to inputs

9.2 The Electro-Magnetic Suspension for a Maglev Vehicle

9.2.1 Introduction

The concept of a magnetically levitated (maglev) vehicle is not new, but recent advances in technology made it possible for maglev vehicles to be successfully implemented. The two most effective suspension techniques are the electrodynamic suspension (EDS) and the electro-magnetic suspension (EMS). EDS systems rely on the fact that a repulsive force is generated when a magnet is moved over a conducting sheet. In order to produce enough force for practical implementation, superconductors are required. The Japanese National Railways have made the most progress with this technology, and have tested a system with a 50 tonne vehicle operating at 300 km/h on a 12.5 km track [Jay88]. Recent discoveries of new superconducting materials may mean that, in the future, this will be the more successful system but, at the present, the EMS system represents the better option.

EMS systems use the attractive force of sets of electromagnets acting upwards towards steel tracks to levitate the vehicle. Japanese Air Lines had a 55 km/h system operating on a 400 m track at Tsukuba Expo 85. In Great Britain, a low velocity 8 tonne commercial passenger-carrying system has been implemented over a 620 m track at Birmingham Airport [Sin87]. The most advanced EMS system is the Transrapid; speeds of 400 km/h have been reached on a 31.5 km test track at Emsland, Germany. A commercial Transrapid system is planned to link Orlando, Florida with Disney World. The link will be 22 km long and the system should be operational by 1996.

9.2.2 The Maglev Vehicle

Recent rapid advances in solid state electronics have meant that electro – magnetically levitated transport systems are now a feasible option. Dynamic control is required for the vertical lift system in order to provide stability to an inherently unstable system, to maintain the airgap, and to ensure the quality of the ride for the passengers. The necessity to maintain the airgap to prevent contact between the vehicle and the guideway means that the vertical control system is a critical system.

The design considered here is an initial design using a linearised one – dimensional model, considering just the vertical heave dynamics. The method could similarly be applied to a more complex three-dimensional model, which would include pitch, roll and yaw dynamics along with lateral and vertical heave dynamics. The model is for a vehicle consisting of a chassis supporting a passenger cabin by means of a secondary suspension consisting of airsprings and hydraulic shock absorbers. The chassis is levitated by means of dc electromagnets under

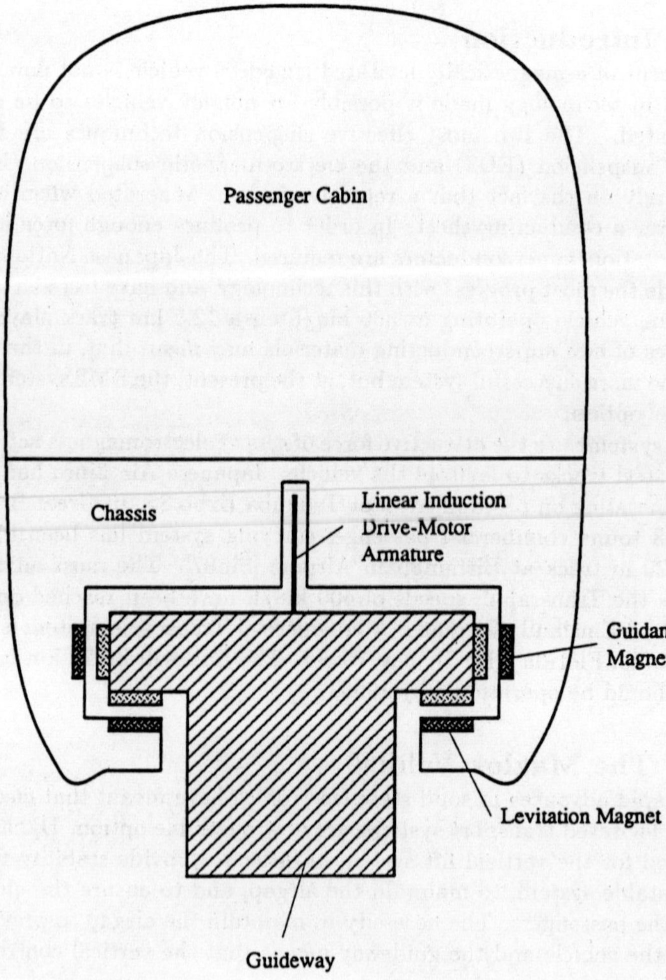

Figure 9.9: Maglev vehicle cross-section

active control providing an attractive force to the guideway. The cross-section of the vehicle showing the configuration is given in Figure 9.9. The vehicle is also equipped with emergency skids in case of power failure.

9.2.3 Theory

The maglev suspension system has three outputs to be bounded, and an input consisting of the sum of two inputs. Accordingly, let

$$w = w_1 + w_2, \tag{9.2.1}$$

with input w_1 belonging to the input space $\mathcal{F}(D)$ as in Definition 8.1 with a corresponding bound D. The second input w_2 is known only to the extent that it belongs to the known input space $\mathcal{F}(A)$ defined as in (9.A.4) with a corresponding bound A. Input space $\mathcal{F}(A)$ is in Definition 9.1 in Appendix 9.A and is characterised by a bound on the second derivative of the input. A relationship to determine the supremum of the output for an input characterised by such a bound is also derived and given by Theorem 9.1 in Appendix 9.A. Conditions for stability are also given.

The outputs are denoted by $v_i(t, w, p)$ for $i = 1, 2, 3$. For all $w_1 \in \mathcal{F}(D)$ and for all $w_2 \in \mathcal{F}(A)$, the outputs are related to the input by

$$v_i(t, w, p) = \int_0^t v_i(t-\tau, h, p) w_1^{(1)}(\tau)\, d\tau + \int_0^t v_i(t-\tau, r, p) w_2^{(2)}(\tau)\, d\tau, \quad i = 1, 2, 3, \tag{9.2.2}$$

where the unit step responses $v_i(t, h, p)$ and the unit ramp responses $v_i(t, r, p)$ are piecewise continuous ($v_i(t, r, p)$ are the integrals of the unit step responses $v_i(t, h, p)$). Following from Zakian [Zak87a] and (8.5.2) and (9.A.6),

$$\phi_{v_i}(p) = D \int_0^\infty |v_i(t, h, p)|\, dt \quad + \quad A \int_0^\infty |v_i(t, r, p)|\, dt, \quad i = 1, 2, 3, \tag{9.2.3}$$

where

$$\phi_{v_i}(p) = \sup \{|v_i(t, w, p)| : t \in \mathbb{R}, w_1 \in \mathcal{F}(D), w_2 \in \mathcal{F}(A)\}. \tag{9.2.4}$$

The design criteria are expressed by

$$\phi_{v_i}(p) \le \varepsilon_i, \quad i = 1, 2, 3, \tag{9.2.5}$$

where ε_i is the largest tolerable value of $|v_i(t, w, p)|$.

In order for

$$\phi_{v_i}(p) < \infty, \quad i = 1, 2, 3, \tag{9.2.6}$$

the following conditions, from (9.A.16) and (8.5.13), are necessary and sufficient:

$$V_i(0, h, p) = 0, \quad \text{for} \quad i = 1, 2, 3, \tag{9.2.7}$$

$$\text{Re}\{\lambda_i(p)\} < 0, \quad \forall \quad \lambda_i(p) \in \Lambda_i(p) \quad \text{for} \quad i = 1, 2, 3, \tag{9.2.8}$$

where $V_i(s, h, p)$ is the Laplace transform of $v_i(t, h, p)$ and $\Lambda_i(p)$ denotes the set of all the finite poles $\lambda_i(p)$ of $V_i(s, h, p)$.

9.2.4 Maglev EMS Model

The one-dimensional design model [KU84, Mül77] considers the vertical movement of the chassis and passenger cabin. The secondary suspension is modelled as a linear spring damper system, and the primary electro-magnetic suspension is described by a linear first order differential equation for the vertical force being derived from the magnet force law and the current/voltage relation. The model configuration is shown in Figure 9.10.

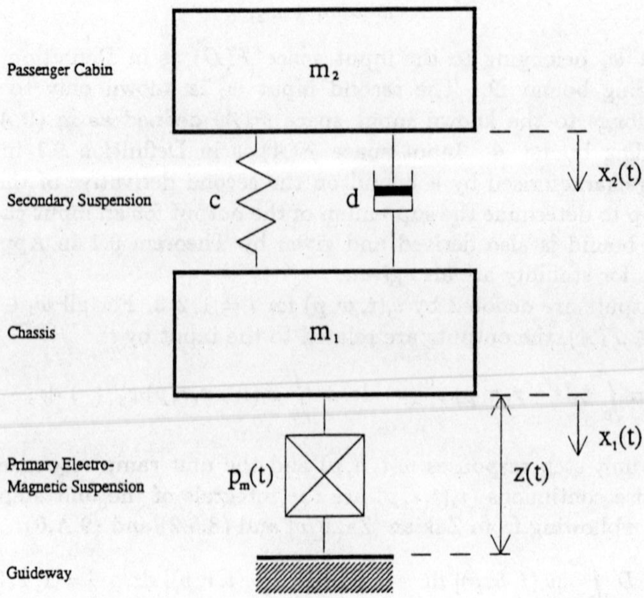

Figure 9.10: Maglev vehicle model configuration

For the electromagnet and fixed track shown in Figure 9.11 [Sin87], the instantaneous electro-magnetic flux linkage between the two bodies through the airgap $z(t)$ is Φ_m. Now, if there is little flux leakage, it can be assumed that $\Phi_m \simeq \Phi_T$ where Φ_T is the flux in the magnet. The instantaneous magnet inductance $L(z,i)$ is then

$$L(z,i) = \frac{N}{i}\Phi_T = \frac{N}{i(t)}\frac{Ni(t)}{R_T}, \tag{9.2.9}$$

where R_T is the reluctance of the entire magnetic circuit, N is the number of turns and $i(t)$ is the current. If the reluctance in the magnetic core is assumed to be negligible compared with the two airgaps (with a total length $2z(t)$), then

$$L(z) = \frac{\mu_0 N^2 A_p}{2z(t)}, \tag{9.2.10}$$

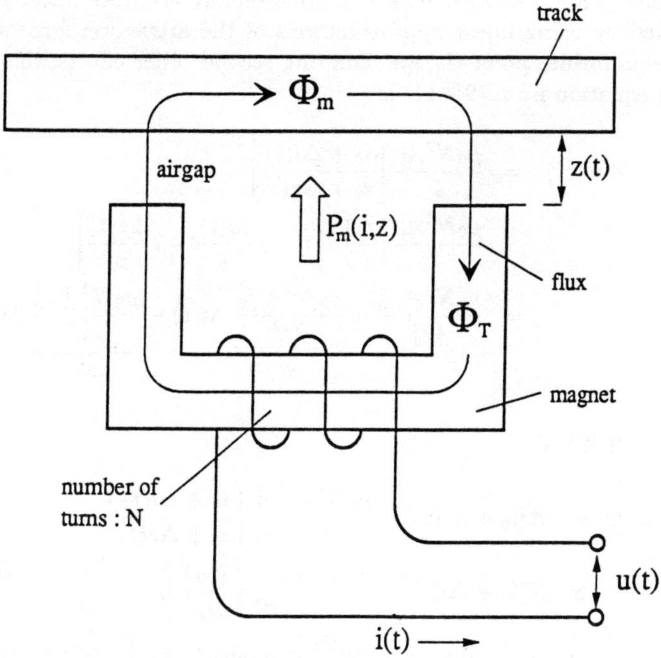

Figure 9.11: Electro-magnetic track configuration

where μ_0 is the permeability of free space and A_p is the pole force area.

The force of attraction p_m at any instant of time is

$$p_m(i,z) = \frac{B^2 A_p}{\mu_0} = \frac{\mu_0 N^2 A_p}{4}\left[\frac{i(t)}{z(t)}\right]^2, \tag{9.2.11}$$

where B is the magnetic flux density. If R is the total resistance of the circuit (including the amplifier output resistance and the magnet winding resistance), then for an instantaneous voltage $u(t)$ across the magnet winding, the excitation current $i(t)$ is controlled by

$$\begin{aligned} u(t) &= Ri(t) + \frac{\mathrm{d}}{\mathrm{d}t}\left[L(z,i)i(t)\right], \\ &= Ri(t) + \frac{\mu_0 N^2 A_p}{2}\frac{\mathrm{d}}{\mathrm{d}t}\left[\frac{i(t)}{z(t)}\right], \\ &= Ri(t) + \frac{\mu_0 N^2 A_p}{2z(t)}\frac{\mathrm{d}i(t)}{\mathrm{d}t} - \frac{\mu_0 N^2 A_p i(t)}{2[z(t)]^2}\frac{\mathrm{d}z(t)}{\mathrm{d}t}. \end{aligned} \tag{9.2.12}$$

Exact analysis of the suspension dynamics would require numerical solution of equations (9.2.11) and (9.2.12), but a reasonably accurate linear model may be obtained by using linear approximations of the attraction force around the nominal equilibrium point (i_0, z_0). Ignoring second order effects, the small perturbation equation from (9.2.11) is

$$\begin{aligned} p_{m_0} + \Delta p_m(t) &= \frac{\mu_0 N^2 A_p}{4}\left[\frac{i_0 + \Delta i(t)}{z_0 + \Delta z(t)}\right]^2, \\ &\simeq \frac{\mu_0 N^2 A_p}{4}\left[\frac{i_0}{z_0}\right]^2\left[1 + 2\frac{\Delta i(t)}{i_0} - 2\frac{\Delta z(t)}{z_0}\right], \\ &= \frac{\mu_0 N^2 A_p i_0^2}{4z_0^2} + \underbrace{\frac{\mu_0 N^2 A_p i_0}{2z_0^2}}_{K_1}\Delta i(t) - \underbrace{\frac{\mu_0 N^2 A_p i_0^2}{2z_0^3}}_{K_2}\Delta z(t), \end{aligned} \tag{9.2.13}$$

and from (9.2.12) is

$$\begin{aligned} u_0 + \Delta u(t) &= R\left[i_0 + \Delta i(t)\right] + \frac{\mu_0 N^2 A_p}{2}\frac{\mathrm{d}}{\mathrm{d}t}\left[\frac{i_0 + \Delta i(t)}{z_0 + \Delta z(t)}\right], \\ &\simeq R\left[i_0 + \Delta i(t)\right] + \frac{\mu_0 N^2 A_p}{2}\frac{\mathrm{d}}{\mathrm{d}t}\left\{\left[\frac{i_0}{z_0}\right]\left[1 + \frac{\Delta i(t)}{i_0} - \frac{\Delta z(t)}{z_0}\right]\right\}, \\ &= Ri_0 + R\Delta i(t) + \underbrace{\frac{\mu_0 N^2 A_p}{2z_0}}_{L}\Delta i^{(1)}(t) - \underbrace{\frac{\mu_0 N^2 A_p i_0}{2z_0^2}}_{K_3}\Delta z^{(1)}(t). \end{aligned} \tag{9.2.14}$$

The chassis has mass m_1 and the passenger cabin has mass m_2. The secondary suspension has a spring constant c and damping constant d. The relationship between the two is assumed linear and satisfies Newton's law:

$$m_1 x_1^{(2)}(t) + d\left[x_1^{(1)}(t) - x_2^{(1)}(t)\right] + c\left[x_1(t) - x_2(t)\right] = p_m(t), \tag{9.2.15}$$

$$m_2 x_2^{(2)}(t) + d\left[x_2^{(1)}(t) - x_1^{(1)}(t)\right] + c\left[x_2(t) - x_1(t)\right] = 0, \tag{9.2.16}$$

where x_1 is the absolute position of the chassis, and x_2 is the absolute position of the passenger cabin.

If small perturbations are assumed, and we take $i(t) = \Delta i(t)$, $z(t) = \Delta z(t)$, $p_m(t) = \Delta p_m(t)$ and $u(t) = \Delta u(t)$, then (9.2.13) becomes

$$p_m(t) = K_1 i(t) - K_2 z(t), \tag{9.2.17}$$

and (9.2.14) becomes

$$u(t) = Ri(t) + Li^{(1)}(t) - K_3 z^{(1)}(t). \tag{9.2.18}$$

The airgap is related to the absolute chassis position by

$$z(t) = w(t) - x_1(t), \tag{9.2.19}$$

where $w(t)$ is the disturbance resulting from variations in the guideway profile.

Typical values of the constants, taken from measurements, are [Mül77]

$$\begin{aligned} K_1 &= 580 \text{ N/A}, \\ K_2 &= 1.3 \times 10^6 \text{ N/m}, \\ K_3 &= 580 \text{ Vs/m}, \\ L &= 0.516 \text{ Vs/A}, \\ R &= 2 \text{ ohms}, \\ m_1 &= 500 \text{ kg}, \\ m_2 &= 500 \text{ kg}, \\ c &= 10^5 \text{ N/m}, \\ d &= 10^4 \text{ Ns/m}. \end{aligned} \tag{9.2.20}$$

The system configuration with the guideway disturbance is shown in Figure 9.12, where the transform functions are, from (9.2.15) to (9.2.19),

$$G_u(s) = \frac{580}{0.516s + 2}, \tag{9.2.21}$$

$$G_1(s) = \frac{s^2 + 20s + 200}{500(s^2 + 40s + 400)}, \tag{9.2.22}$$

$$G_2(s) = \frac{(s + 10)}{25(s^2 + 40s + 400)}, \tag{9.2.23}$$

$$G_3(s) = \frac{334400s + 2600000}{0.516s + 2}. \tag{9.2.24}$$

9.2.5 The Input Space

The major disturbance to the system is from variations in the guideway height. These variations arise from two sources; the geometry of the guideway can thus be represented by

$$w(t) = w_1(t) + w_2(t), \tag{9.2.25}$$

where $w_2(t)$ represents the guideway variations due to gradients in the terrain, and $w_1(t)$ are irregularities in the guideway, caused mainly by sagging between the guideway supports as well as by maladjustments in the rail and thermal effects. The following bounds for the guideway have been suggested for a 140 m/s vehicle [Mül77] :

$$\sup\left\{\left|w_1^{(1)}(t)\right| : t \in \mathbb{R}^+\right\} = 0.030 \text{ m/s} \tag{9.2.26}$$

and

$$\sup\left\{\left|w_2^{(2)}(t)\right| : t \in \mathbb{R}^+\right\} = 0.100 \text{ m/s}^2. \tag{9.2.27}$$

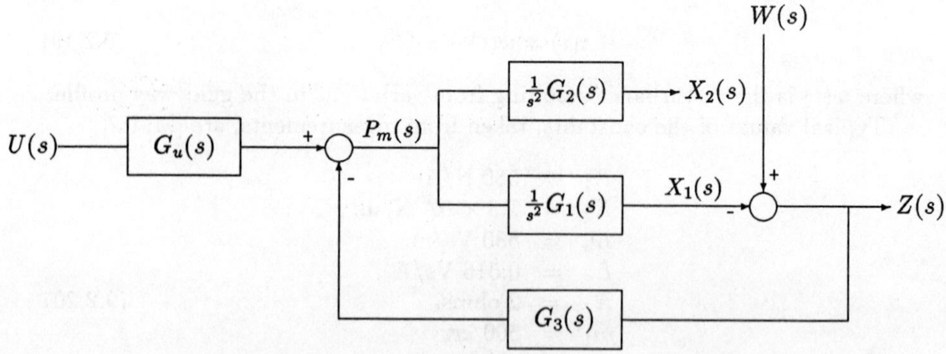

Figure 9.12: System configuration with guideway disturbance

Thus from (8.2.3) and (9.A.4), the two input spaces are characterised by

$$D = 0.030 \text{ m/s}, \tag{9.2.28}$$

and

$$A = 0.100 \text{ m/s}^2. \tag{9.2.29}$$

The rate bound on w_1 reflects the quality of the guideway, and the acceleration bound on w_2 represents a restriction on the suddenness with which a gradient begins.

9.2.6 Design Criteria

The primary purpose of the vertical control system is to provide stability to an otherwise unstable system. An electromagnet excited by a constant voltage will either clamp to the rail or fall because the attractive force decreases with increasing airgap. Once stability has been ensured by the control scheme there are two important outputs which have to be considered in determining the effectiveness of the control system:

i) It is desirable to avoid contact between the guideway and the skids in order that wear and possible damage to the guideway and skids is minimised, and to avoid unnecessary vibration, friction and noise. In addition, frequent contact leads to a deterioration in the ride quality. This means that the control system is critical, in that the minimum permissible airgap has a definite bound. The allowed airgap is between 7 mm and 17 mm which means a nominal gap of 12 mm. A minimal gap of 7 mm assures sufficient safety margin allowing for the failure of single magnets [KU84]. The nominal airgap cannot be too large as lifting capacity decreases with growing airgap, and thus the nominal airgap must be limited in order to obtain reasonable magnet weights and power consumption. Thus, the maximum error permitted between the airgap and the nominal airgap is 5 mm.

ii) The ride comfort can be determined by the amount of vertical acceleration experienced by the passengers. Thus it is important that the maximum acceleration in either direction is not too great. Typically, this should not exceed 0.5 m/s^2 [KU84]. This second criterion is important, but is not rigid and may be increased to as much as 1.0 m/s^2 [Arm84].

iii) An additional criterion is in the amount of control effort required; the required control voltage should be within -600 V and + 600 V [KU84].

The design criteria (9.2.5) are rewritten

$$\phi_{v_i} \le \varepsilon_i, \quad i = 1, 2, 3, \tag{9.2.30}$$

where

$$v_1 = e, \tag{9.2.31}$$
$$v_2 = x_2^{(2)}, \tag{9.2.32}$$
$$v_3 = u, \tag{9.2.33}$$

and

$$\varepsilon_1 = 5 \text{ mm}, \tag{9.2.34}$$
$$\varepsilon_2 = 500 \text{ mm/s}^2, \tag{9.2.35}$$
$$\varepsilon_3 = 600 \text{ V}. \tag{9.2.36}$$

9.2.7 The Design

Thus with the design criteria specified, and the plant and the input spaces determined, relationship (9.2.3) was used with the 'Criteria' package implemented on a Sun Workstation to determine a set of controllers that would satisfy the criteria.

An inductive sensor is used to measure the gap width z, and an accelerometer for the chassis acceleration $x_1^{(2)}$. These two signals are used to implement a feedback control scheme as shown in Figure 9.13.

From Figure 9.13, the closed-loop transfer function $V_1(s,p)$ relating the airgap error $E(s)$ to the input $W(s)$ is

$$V_1(s,p) = -\frac{1 + K_1(s,p)G_1(s)G_u(s)}{1 + K_1(s,p)G_1(s)G_u(s) - \frac{1}{s^2}G_1(s)G_3(s) - \frac{1}{s^2}K_z(s,p)G_1(s)G_u(s)}, \tag{9.2.37}$$

and the closed-loop transfer function $V_2(s,p)$ relating the carriage acceleration $X_2^{(2)}(s)$ to the input $W(s)$ is

$$V_2(s,p) = \frac{-G_2(s)\,(G_3(s) + K_z(s,p)G_u(s))}{1 + K_1(s,p)G_1(s)G_u(s) - \frac{1}{s^2}G_1(s)G_3(s) - \frac{1}{s^2}K_z(s,p)G_1(s)G_u(s)}, \tag{9.2.38}$$

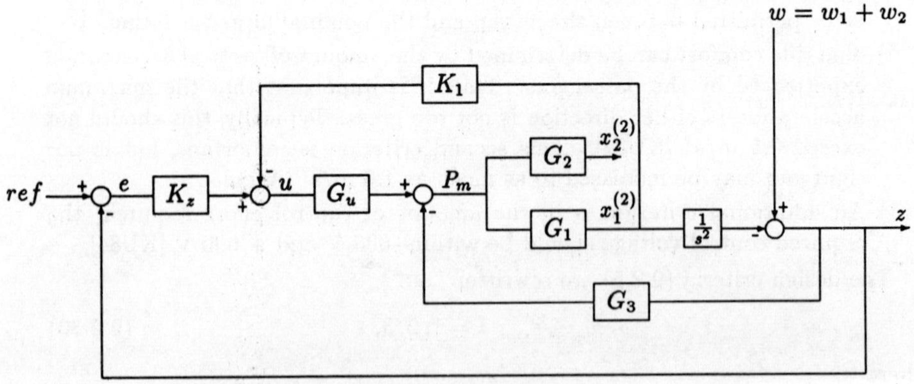

Figure 9.13: Maglev vehicle suspension control scheme

and the closed-loop transfer function $V_3(s,p)$ relating the control voltage $U(s)$ to the input $W(s)$ is

$$V_3(s,p) = \frac{K_1(s,p)G_1(s)G_3(s) - K_z(s,p)}{1 + K_1(s,p)G_1(s)G_u(s) - \frac{1}{s^2}G_1(s)G_3(s) - \frac{1}{s^2}K_z(s,p)G_1(s)G_u(s)}. \tag{9.2.39}$$

The presence of a double integrator in the plant means that condition (9.2.7) is satisfied, so the first stage was to satisfy (9.2.8), i.e. stabilise the system, and this was achieved after only a few iterations of the MBP. The next stage was to satisfy (9.2.30). After a number of iterations, the criteria (9.2.30) had not been satisfied, so the bound ε_2 on the passenger acceleration $x_2^{(2)}$ was relaxed in stages until

$$\varepsilon_1 = 5 \text{ mm}, \tag{9.2.40}$$
$$\varepsilon_2 = 800 \text{ mm/s}^2, \tag{9.2.41}$$
$$\varepsilon_3 = 600 \text{ V}. \tag{9.2.42}$$

With these bounds the criteria (9.2.30) are satisfied by

$$K_z = -1 \times 10^6 \frac{1.7s^2 + 10.5s + 31.5}{0.05s^2 + 20s + 1}, \tag{9.2.43}$$
$$K_1 = 1 \times 10^6 \frac{2.22s^2 + 16.0s + 0.73}{648s^2 + 61s + 1}, \tag{9.2.44}$$

giving

$$\phi_{v_1} = 4.81 \text{ mm}, \tag{9.2.45}$$

$$\phi_{v_2} = 800 \text{ mm/s}^2, \tag{9.2.46}$$
$$\phi_{v_3} = 238 \text{ V}. \tag{9.2.47}$$

The bound ε_2 is not rigid; the consequence of its being increased is a lower level of passenger comfort. However, ϕ_{v_2} represents the greatest possible acceleration of the passenger carriage, and may only occur rarely, hence this design is acceptable.

A typical guideway disturbance is shown in Figure 9.14. Input $w_1(t)$ is a constant gradient of 30 mm/s followed by a sinusoid with a maximum gradient of 30mm/s, and $w_2(t)$ is a curve with $w_2^{(2)}(t) = 100$ mm/s^2 followed by a constant gradient. The summed input $w(t)$ therefore represents a severe test of the system. Responses to this input are shown in Figures 9.15, 9.16 and 9.17, and it can be seen from these that the bounds (9.2.40) to (9.2.42) are not exceeded.

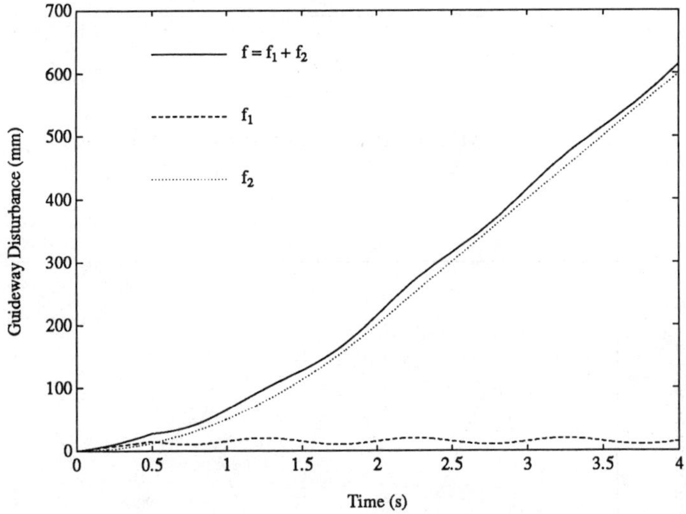

Figure 9.14: Typical input

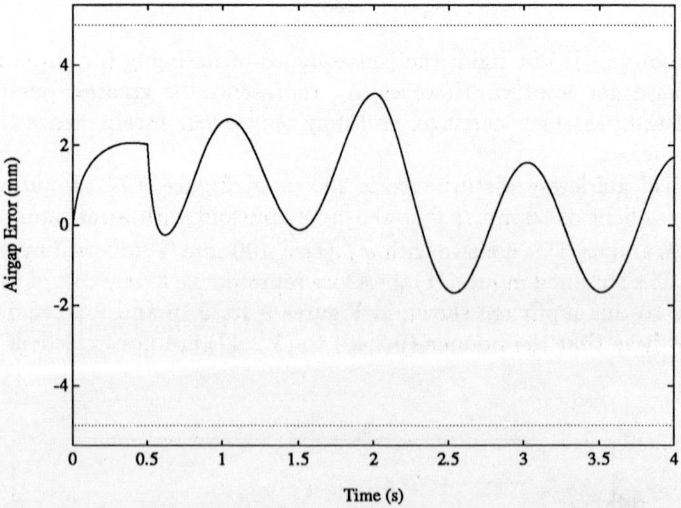

Figure 9.15: Airgap error response to typical input

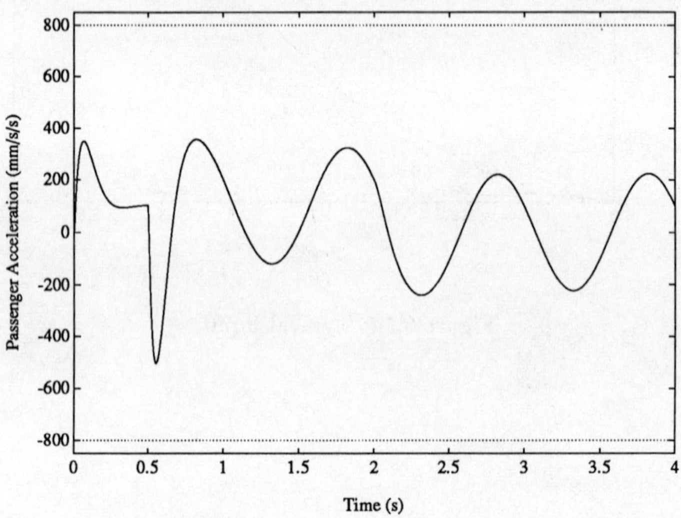

Figure 9.16: Passenger carriage acceleration response to typical input

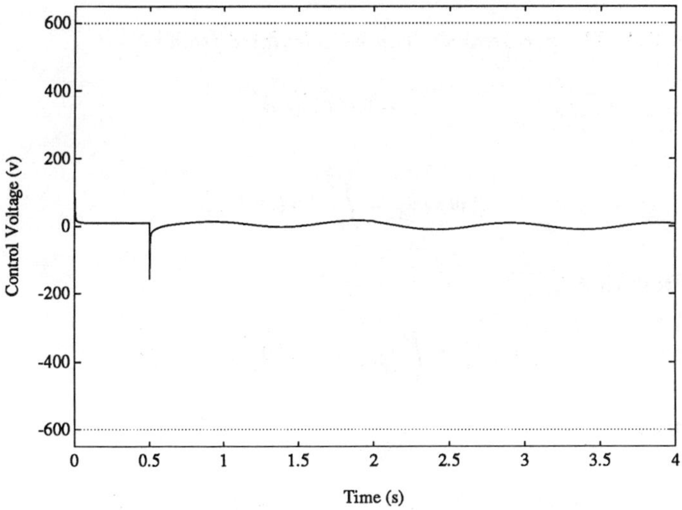

Figure 9.17: Control voltage response to typical input

9.A Appendix - Input Space $\mathcal{F}(A)$

Definition 9.1 *The input space $\mathcal{F}(A)$ is the class of all functions $w : \mathbb{R} \to \mathbb{R}$ such that*

$$w^{(2)}(t) = \frac{\mathrm{d}^2 w(t)}{\mathrm{d}t^2} \tag{9.A.1}$$

is piecewise continuous and where

$$w(t) = 0 \quad \forall \quad t \leq 0, \tag{9.A.2}$$

and

$$w^{(1)}(t) = 0 \qquad \text{for} \qquad t = 0, \tag{9.A.3}$$

and there is also a real number A where $0 < A < \infty$, such that

$$A = \sup \left\{ \left| w^{(2)}(t) \right| : t \in \mathbb{R}^+, w \in \mathcal{F}(A) \right\}. \tag{9.A.4}$$

The output is $v(w)$, the value of the output at time t is denoted by $v(t, w)$. It is assumed that the input-output function $w \mapsto v(w)$ is causal, linear and time-invariant and that for all $w \in \mathcal{F}(A)$ the output is related to the input by the convolution integral given by (8.5.1) where the unit impulse step $v(t, h)$ is piecewise continuous. After integration by parts, (8.5.1) can be rearranged to

$$v(t, w) = \int_0^t v(t - \tau, r) w^{(2)}(\tau) \, \mathrm{d}\tau, \tag{9.A.5}$$

where $v(r)$ is the unit ramp response, that being the integral of the step response.

Theorem 9.1 *The performance can be calculated from*

$$\phi_v = \| v(r) \|_1 A, \tag{9.A.6}$$

where

$$\| v(r) \|_1 = \int_0^\infty |v(t,r)| \, dt. \tag{9.A.7}$$

Proof: From (9.A.5)

$$|v(t,w)| \leq \int_0^t |v(\tau,r)| \left| w^{(2)}(t-\tau)) \right| d\tau. \tag{9.A.8}$$

From (9.A.4)

$$\sup \{|v(t,w)| : w \in \mathcal{F}(A)\} \leq A \int_0^t |v(\tau,r)| \, d\tau, \tag{9.A.9}$$

which gives

$$\sup \left\{ |v(t,w)| : t \in \mathbb{R}^+, w \in \mathcal{F}(A) \right\} \leq A \int_0^\infty |v(\tau,r)| \, d\tau. \tag{9.A.10}$$

Now consider $w_t \in \mathcal{F}(A)$ where

$$w_t^{(2)}(t-\tau) = \begin{cases} A \text{ sgn } v(\tau,r), & \text{for} \quad 0 \leq \tau, \\ 0 & \text{for} \quad \tau < 0, \end{cases} \tag{9.A.11}$$

which from (9.A.5) gives

$$v(t,w_t) = A \int_0^t |v(\tau,r)| \, d\tau. \tag{9.A.12}$$

So, as $t \to \infty$,

$$v(\infty,w_t) = A \int_0^\infty |v(\tau,r)| \, d\tau. \tag{9.A.13}$$

So from (9.A.10) and (9.A.13)

$$\phi_v = \| v(r) \|_1 A, \tag{9.A.14}$$

and the theorem is proved. □□

Stability

Theorem 9.2 *In order for*

$$\phi_v < \infty, \tag{9.A.15}$$

the following conditions are necessary and sufficient:

$$\begin{array}{ll} a) & V(0,h) = 0, \\ b) & \mathrm{Re}\{\lambda\} < 0 \quad \forall \quad \lambda \in \Lambda, \end{array} \tag{9.A.16}$$

where $V(s,h)$ is the Laplace transform of $v(t,h)$ and Λ denotes the set of all the finite poles λ of $V(s,h)$.

Proof: The proof follows from (9.A.6) and (8.5.13). □□

Chapter 10

Performance in Sampled-Data Systems

10.1 Introduction

A performance index suitable for critical control system design in the continuous time domain is defined in Section 8.3. The lack of performance ϕ_e is defined as the least upper bound of the absolute error that occurs over all time and all inputs:

$$\phi_e = \sup\{|e(t,w)| : t \in \mathbb{R}, w \in \mathcal{F}\}, \qquad (10.1.1)$$

where $e(t,w)$ is the system error, t is time and $w \in \mathcal{F}$ is the system input, which belongs to the known function space of inputs $\mathcal{F}$. In Chapter 8, expressions to determine ϕ_e for linear time-invariant continuous control systems were given, and, in particular, an expression (8.5.2) was derived to determine ϕ_e for systems where the input space $\mathcal{F}(D)$ is characterised by a uniform bound on the derivative of the input. However, expression (8.5.2) is not applicable to sampled-data control systems. This chapter shows how ϕ_e can be determined for certain sampled-data control systems.

The calculation of ϕ_e is less simple for sampled-data systems than for continuous time systems. However, it is relatively simple to calculate the lack of performance ϕ_e^* at the sampling instances rather than over all time, where

$$\phi_e^* = \sup\{|e^*(k,w^*)| : k \in \mathbb{N}^+ , \quad w^* \in \mathbb{D}(0,\delta)\}, \qquad (10.1.2)$$

where $e^*(k,w^*)$ is the system error at the sampling instances, k is the unit of discrete time and $w^* \in \mathbb{D}(0,\delta)$ is the system input, which belongs to the known function space of inputs $\mathbb{D}(0,\delta)$. An expression is derived in Section 10.2 which shows how ϕ_e^* can be determined for any linear sampled-data system where the input space $\mathbb{D}(0,\delta)$ is characterised by a uniform bound on the input increments. This expression is analogous to Theorem 8.1 derived for continuous time systems.

The error between the sampling instances may exceed the error at the sampling instances; in such cases ϕ_e is greater than ϕ_e^*. Section 10.3 shows how ϕ_e

can be determined for the sampled-data control system shown in Figure 10.1 where $w \in \mathcal{F}(D)$. The relationship between $\mathcal{F}(D)$ and $\mathbb{D}(0,\delta)$, and hence the relationship between ϕ_e and ϕ_e^* is given in Section 10.4

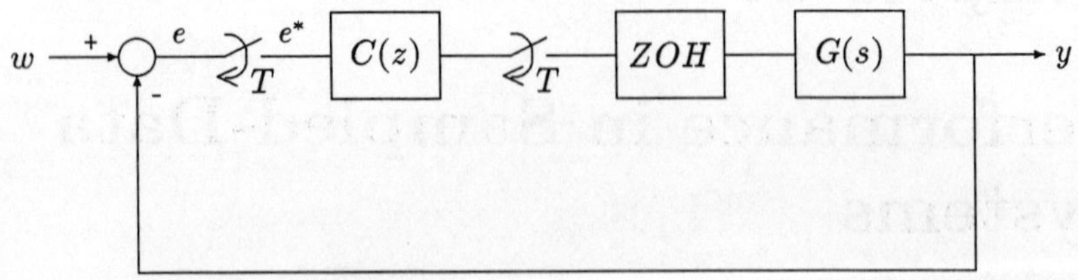

Figure 10.1: Sampled-data control system configuration

Section 10.5 outlines the procedure for design using the method of inequalities with ϕ_e as the performance index, and Section 10.6 shows the design of the control system for the pointing mechanism of a satellite earth scanning antenna.

10.2 Performance at Sampling Points

The input $w^*(k)$ is known only to the extent that it belongs to the known function space $\mathbb{D}(0,\delta)$ which is defined as the class of all functions $w^* : \mathbb{N} \to \mathbb{R}$ such that

$$w^*(k) = 0 \quad \forall \quad k = -1, -2, \dots, \quad w^* \in \mathbb{D}(0,\delta), \tag{10.2.1}$$

and there is a finite positive number δ such that

$$\delta = \sup\{|\Delta(k)| : k \in \mathbb{N}^+, \quad w^* \in \mathbb{D}(0,\delta)\}, \tag{10.2.2}$$

where

$$\Delta(k) = w^*(k) - w^*(k-1). \tag{10.2.3}$$

The response of the error at the kth unit of discrete time to the input w^* is denoted by $e^*(k, w^*)$. For all $w^* \in \mathbb{D}(0,\delta)$ the output is related to the input by the convolution summation

$$e^*(k, w^*) = \sum_{n=0}^{k} e^*(k-n)w^*(n). \tag{10.2.4}$$

The response $e^*(k-n)$ is simply the weighting sequence $e^*(k)$ delayed n units of discrete time, where $e^*(k)$ is the response to the Kronecker delta sequence applied at $k = 0$. The unit step response $e^*(k,h)$ is the response to the unit step $h(k)$ where

$$h(k) = \begin{cases} 0 & \text{for} \quad k < 0, \\ 1 & \text{for} \quad k \geq 0. \end{cases} \tag{10.2.5}$$

Thus

$$e^*(k,h) = \sum_{n=0}^{k} e^*(k-n) = \sum_{n=0}^{k} e^*(n), \tag{10.2.6}$$

so the input output rule can be expressed as

$$\begin{aligned} e^*(k,w^*) &= \sum_{n=0}^{k} e^*(k-n)w^*(n), \\ &= \sum_{n=0}^{k} \left\{ \sum_{i=0}^{k-n} e^*(i) - \sum_{i=0}^{k-n-1} e^*(i) \right\} w^*(n), \\ &= \sum_{n=0}^{k} \left\{ e^*(k-n,h)w^*(n) - e^*(k-n-1,h)w^*(n) \right\}, \\ &= \sum_{n=0}^{k} e^*(k-n,h) \left\{ w^*(n) - w^*(n-1) \right\}, \\ &= \sum_{n=0}^{k} e^*(k-n,h)\Delta(k). \end{aligned} \tag{10.2.7}$$

A bound ϕ_e^* on the error is now stated as

$$\phi_e^* = \sup \left\{ |e^*(k,w^*)| : k \in \mathbb{N}^+ , \quad w^* \in \mathbb{D}(0,\delta) \right\}. \tag{10.2.8}$$

ϕ_e^* is the greatest error as k ranges over the space $\mathbb{N}$ and the input ranges over the input space $\mathbb{D}(0,\delta)$.

Theorem 10.1 *The following result*

$$\phi_e^* = \delta \parallel e^*(h) \parallel_1, \tag{10.2.9}$$

provides a simple formula for computing ϕ_e^* *where*

$$\parallel e^*(h) \parallel_1 = \sum_{n=0}^{\infty} |e^*(n,h)|. \tag{10.2.10}$$

Proof: Now

$$\phi_e^* = \sup\{|e^*(k,w^*)| : k \in \mathbb{N}^+ , \quad w^* \in \mathbb{D}(0,\delta)\}. \tag{10.2.11}$$

Since $w^*(k) = 0$ for $k = -1, -2, \ldots$, it follows that, from (10.2.7)

$$|e^*(k,w^*)| \leq \sum_{n=0}^{k} |e^*(k-n,h)| \, |w^*(n) - w^*(n-1)|, \tag{10.2.12}$$

$$\sup_{w^* \in \mathbb{D}(0,\delta)} |e^*(k,w^*)| \leq \delta \sum_{n=0}^{k} |e^*(k-n,h)|\,, \tag{10.2.13}$$

$$\sup_{w^* \in \mathbb{D}(0,\delta)} \sup_{k \in \mathbb{N}} |e^*(k,w^*)| \leq \delta \sum_{n=0}^{\infty} |e^*(n,h)|\,, \tag{10.2.14}$$

$$\phi_e^* \leq \delta \parallel e(h) \parallel_1. \tag{10.2.15}$$

Now consider an input sequence $w_t^* \in \mathbb{D}(0,\delta)$ where

$$w_t^*(n) - w_t^*(n-1) = \begin{cases} \delta \text{ sgn } e^*(k-n) & \text{for} \quad 0 \leq n \leq k, \\ 0 & \text{for} \quad n < 0, \end{cases} \tag{10.2.16}$$

which from (10.2.7) gives

$$\begin{aligned} e^*(k,w_t^*) &= \delta \sum_{n=0}^{k} |e^*(k-n,h)|, \\ &= \delta \sum_{n=0}^{k} |e^*(n,h)|. \end{aligned} \tag{10.2.17}$$

So, as $k \to \infty$,

$$e^*(\infty, w_t^*) = \delta \parallel e^*(h) \parallel_1. \tag{10.2.18}$$

So from (10.2.15) and (10.2.18)

$$\phi_e^* = \delta \parallel e^*(h) \parallel_1, \tag{10.2.19}$$

and the theorem is proved. □□

This theorem is analogous to that for the continuous time case given by Theorem 8.1.

10.3 Performance Over All Time

The limitation of Theorem 10.1 is that it gives the supremum of the output at the sampling points only. This value may be exceeded at times between the sampling points, so a means of calculating the supremum of the error at all times ϕ_e, as defined by (10.1.1) is required.

Therefore the system is considered as a continuous time system, with continuous inputs and outputs. However, the process of sampling means that a digital control scheme which is composed of linear time-invariant (LTI) components and fixed parameter digital controllers is a linear time-variant (LTV) system when considered as a continuous time system.

Consider the system shown in Figure 10.2. An input which occurs at a time between samples will not be seen by the plant until the next sample, so the plant cannot respond until the sampling instant. The relationship between the system

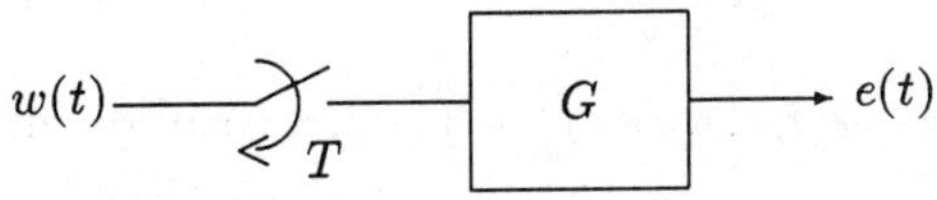

Figure 10.2: A sampled-data system

input and output is thus not independent of time, therefore the system cannot be time-invariant.

For a causal LTV system, the error at any time is related to the input by the linear time-variant convolution [ZD63]

$$e(t,w) = \int_0^t e(t,\tau,\delta)w(\tau)\ d\tau, \tag{10.3.1}$$

where $w(t)$ is the input and $e(t,\tau,\delta)$ is the response at time t to a unit impulse $\delta(t-\tau)$ applied at time τ. This LTV convolution (10.3.1) differs from the LTI convolution,

$$e(t,w) = \int_0^t e(t-\tau,\delta)w(\tau)\ d\tau, \tag{10.3.2}$$

in that (10.3.2) is dependent upon the interval $t-\tau$, whereas (10.3.1) is a function of the initial and final times.

Thus in order be able to calculate ϕ_e, an expression which gives the supremum of the error at any particular time for all LTV systems is first derived. From this, an expression from which ϕ_e can be calculated is derived for the particular sampled-data control scheme shown in Figure 10.1.

The input is known to the extent that it belongs to the known function space $\mathcal{F}(D)$ which is defined by Definition 8.1.

Let $e(t,\tau,h)$ denote the response at time t to a unit step $h(t-\tau)$ where $h(t-\tau)=0$ for $t \le \tau$ and $h(t-\tau)=1$ for $t > \tau$. From (10.3.1)

$$e(t,\tau,h) = \int_\tau^t e(t,\tau',\delta)\ d\tau', \tag{10.3.3}$$

which can be expressed as

$$\begin{aligned} e(t,\tau,h) &= \int_0^t e(t,\tau',\delta)\ d\tau' - \int_0^\tau e(t,\tau',\delta)\ d\tau', \\ &= e(t,0,h) - \int_0^\tau e(t,\tau',\delta)\ d\tau', \end{aligned} \tag{10.3.4}$$

and differentiating gives

$$\frac{de(t,\tau,h)}{d\tau} = -e(t,\tau,\delta). \tag{10.3.5}$$

Integrating (10.3.1) by parts and substituting (10.3.5) gives

$$e(t,w) = -e(t,\tau,h)w(\tau)\Big|_0^t + \int_0^t e(t,\tau,h)w^{(1)}(\tau)\ \mathrm{d}\tau, \tag{10.3.6}$$

and since $w(0) = 0$ and $e(t,t,h) = 0$,

$$e(t,w) = \int_0^t e(t,\tau,h)w^{(1)}(\tau)\ \mathrm{d}\tau. \tag{10.3.7}$$

Lemma 10.1 *The supremum of the error of a causal linear time-variant system at a specific time t is given by*

$$\sup_{w\in\mathcal{F}(D)} |e(t,w)| = D\int_0^t |e(t,\tau,h)|\ \mathrm{d}\tau. \tag{10.3.8}$$

Proof: From (10.3.7) and (8.2.3)

$$\sup_{w\in\mathcal{F}(D)} |e(t,w)| \leq D\int_0^t |e(t,\tau,h)|\ \mathrm{d}\tau. \tag{10.3.9}$$

For time $t \geq 0$ there exists an $w_t \in \mathcal{F}(D)$ defined by

$$w_t^{(1)}(\tau) = \begin{cases} D \text{ sgn } e(t,\tau,h) & \text{for} \quad 0 \leq \tau \leq t, \\ 0 & \text{for} \quad \tau < 0, \end{cases} \tag{10.3.10}$$

which from (10.3.7) gives

$$\begin{aligned} e(t,w_t) &= \int_0^t e(t,\tau,h) \text{ sgn } e(t,\tau,h)\ \mathrm{d}\tau, \\ &= D\int_0^t |e(t,\tau,h)|\ \mathrm{d}\tau. \end{aligned} \tag{10.3.11}$$

So from (10.3.9) and (10.3.11), the lemma is proved. □□

Theorem 10.2 *For the digital control scheme shown in Figure 10.1, for all $k \in \mathbb{N}^+$ and $w \in \mathcal{F}(D)$ with a given sampling interval T, the supremum of the error at the particular times $kT + t_s$ where $0 \leq t_s < T$, namely*

$$\tilde{e}(t_s) = \sup\left\{|e(kT+t_s,w)| : k \in \mathbb{N}^+ \backslash 0\,, \quad w \in \mathcal{F}(D)\right\}, \tag{10.3.12}$$

is given by

$$\tilde{e}(t_s) = D\left(t_s + T\sum_{k=0}^{\infty} |e(kT+t_s,0,h)|\right). \tag{10.3.13}$$

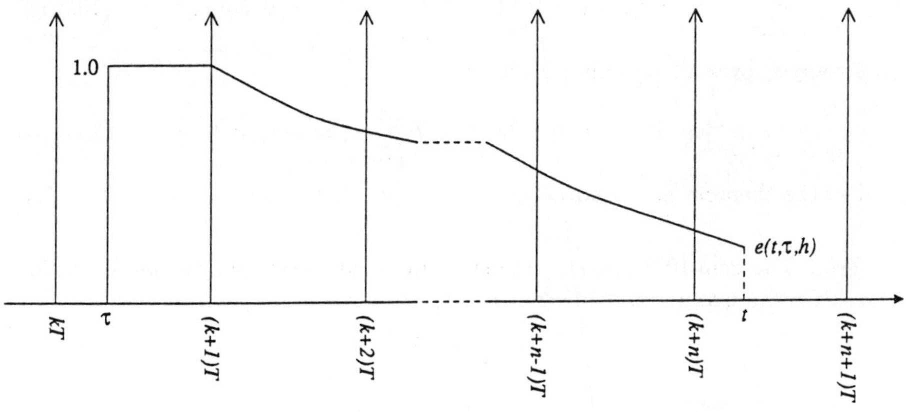

Figure 10.3: Typical error step response for the sampled-data system

Proof: For the control scheme in Figure 10.1, because the input is sampled, $e(t,\tau,h)$ remains constant at a time t for the starting time τ ranging between sampling points. In addition, the initial error is unity, and remains so until the next sampling instant. A typical error step response is shown in Figure 10.3.

So, for $kT \le \tau < (k+1)T$,

$$e(t,\tau,h) = \begin{cases} 0 & \text{for} \quad t \le \tau, \\ 1 & \text{for} \quad \tau < t \le (k+1)T, \\ e(t,kT,h) & \text{for} \quad t > (k+1)T, \end{cases} \tag{10.3.14}$$

where $k = 0, 1, 2, \ldots$. Let $nT < t \le (n+1)T$, and so from (10.3.8)

$$\begin{aligned} \sup_{w \in \mathcal{F}(D)} |e(t,w)| &= D\big\{T\,|e(t,0,h)| + T\,|e(t,T,h)| + T\,|e(t,2T,h)| + \cdots \\ &\quad \cdots + T\,|e(t,(n-1)T,h)| + T\,|e(t,nT,h)| + t_s\big\}, \end{aligned} \tag{10.3.15}$$

which is

$$\sup_{w \in \mathcal{F}(D)} |e(t,w)| = D\left(t_s + T\sum_{k=0}^{n} |e\,(t,kT,h)|\right), \tag{10.3.16}$$

where $n = 0, 1, 2, \ldots$. Let $t = nT + t_s$, then

$$\sup_{w \in \mathcal{F}(D)} |e(nT + t_s, w)| = D\left(t_s + T\sum_{k=0}^{n} |e\,(nT + t_s, kT, h)|\right). \tag{10.3.17}$$

Because of the periodic nature of the system,

$$e(t,\tau,h) = e(t + kT, \tau + kT, h), \qquad k = 0, 1, 2, \ldots, \tag{10.3.18}$$

and so from (10.3.17),

$$\sup_{w\in\mathcal{F}(D)} |e(nT+t_s,w)| = D\left(t_s + T\sum_{k=0}^{n}|e(kT+t_s,0,h)|\right), \qquad (10.3.19)$$

so for any t_s over all sampling instances.

$$\sup_{k\in\mathrm{N}^+}\ \sup_{w\in\mathcal{F}(D)} |e(nT+t_s,w)| = D\left(t_s + T\sum_{k=0}^{\infty}|e(kT+t_s,0,h)|\right), \qquad (10.3.20)$$

and so the theorem is proved. □□

From Theorem 10.2, it is easy to determine ϕ_e numerically by a one-dimension search for the maximum value of $\tilde{e}$ over t_s, so

$$\phi_e = \sup_{0\leq t_s<T} \tilde{e}(t_s). \qquad (10.3.21)$$

10.4 The Relationship Between $\mathcal{F}(D)$ and $\mathbb{D}(0,\delta)$

The input $w^*(k) \in \mathbb{D}(0,\delta)$ represents a periodic sample of an input $w(t) \in \mathcal{F}(D)$. If the sampling period is T, then from (10.2.2) and (8.2.3) then

$$\delta = DT. \qquad (10.4.1)$$

Thus from (10.3.13) for $t_s = 0$, (i.e. at the sampling instant)

$$\tilde{e}(0) = DT\sum_{k=0}^{\infty}|e(kT,0,h)|, \qquad (10.4.2)$$

which is equivalent to (10.2.9).

10.5 Design Procedure

The design problem can be set in the form of the method of inequalities

$$\phi_e(p) \leq \varepsilon, \qquad (10.5.1)$$

where

$$\phi_e(p) = \sup\left\{|e(t,w,p)| : t\in\mathbb{R}, w\in\mathcal{F}(D)\right\}, \qquad (10.5.2)$$

and ε is the maximum tolerable value of the system error e, and $p \in \mathbb{R}^N$ is the set of design parameters, in this case representing the parameters of the controller $C(z,p)$. If inequality (10.5.1) holds, then the performance is guaranteed and the design is complete. The inequality (10.5.1) may be solved by numerical methods such as the moving boundaries process (MBP) described in Section (3.2).

Expression (10.3.21) is more computationally expensive than (10.2.9). Therefore, to reduce the computational effort, the design procedure is conducted in two stages:

i) Use the MBP to solve the amended design problem

$$\phi_e^*(p) \le \varepsilon, \tag{10.5.3}$$

where

$$\phi_e^*(p) = \sup\left\{|e^*(k, w^*, p)| : k \in \mathbb{N}^+, \quad w^* \in \mathbb{D}(0, \delta)\right\}, \tag{10.5.4}$$

using (10.2.9). Calculate $\phi_e(p)$, and if $\phi_e(p) > \varepsilon$ then go on to the second stage.

ii) Solve the design problem (10.5.1) using the MBP in conjunction with (10.3.21).

It should be noted that, from (10.3.14), $e(T, 0, h) = 1$. So, from (10.4.2),

$$\tilde{e}(0) \ge DT, \tag{10.5.5}$$

and hence from (10.3.21)

$$\phi_e \ge DT. \tag{10.5.6}$$

This means that ϕ_e cannot be made less than $D \times T$. It also gives a means of choosing the sampling period T, from (10.5.6) and (10.5.1)

$$T \le \frac{\varepsilon}{D}. \tag{10.5.7}$$

10.6 Design of an Earth Scanning Satellite Antenna Control System

An earth scanning antenna on a space satellite is to be used to collect information on temperature and humidity using sensors which pick up microwave radiation emitted naturally from the earth and its atmosphere. The antenna is required to follow a variety of position reference profiles which have a maximum reference velocity of 1.0 rad/s. The design requirement is that the positional error be less than 1.2 degrees. To keep down hardware costs, a microprocessor based servo control system is used to drive the pointing mechanism.

Figure 10.4 shows a block diagram of a linear model of the servo system dynamics, where:

$u(t)$ Voltage applied across dc servo motor,
$v_e(t)$ Net voltage across motor armature,
$i_a(t)$ Current in motor armature,
$\tau_m(t)$ Torque generated by motor at motor shaft,
$\omega(t)$ Angular velocity of motor shaft,
$y(t)$ Angular position of antenna,
$v_b(t)$ Back EMF generated by the motor,

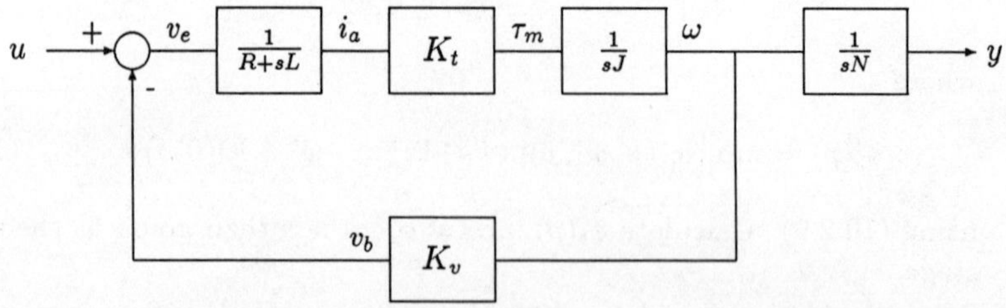

Figure 10.4: Linear model of antenna servo

K_t	Motor torque coefficient,
K_v	Motor back EMF constant (incorporating effect of viscous damping),
R	Resistance of the motor armature winding,
L	Inductance of the motor armature winding,
J	Total mechanism inertia referred to the motor shaft output,
N	Gearbox reduction.

The system dynamics are combined into a transfer function $G(s)$, which relates the angular displacement of the motor output shaft to the voltage applied at the motor. The antenna is assumed stiff, and non-linearities such as stiction and gearbox backlash are ignored.

$$G(s) = \frac{K_t/N}{s(JLs^2 + JRs + K_tK_v)}. \tag{10.6.1}$$

Voltage to the motor is applied by a digital controller $C(z,p)$ via a zero-order hold unit. Input to the controller consists of the error $e(t)$ between the reference position profile signal $w(t)$ and the antenna positional output $y(t)$. It is assumed that both $w(t)$ and $y(t)$ are scaled in radians. The reference signals of the position profiles are obtained from look-up tables, and the angular position of the antenna is measured by an inductosyn, which is sampled by the digital controller with a sampling period T. The discretised reference input profiles have previously been sampled from the desired continuous time profiles. Quantitisation effects of the signal measurements are ignored. The control scheme is thus in the form of Figure 10.1.

The position reference profiles have a maximum reference velocity of 1.0 rad/s, so the input space bound D in Definition 8.1 is $D = 1.0$. The input space bound δ is determined from (10.4.1). The design requirement is

$$\varepsilon = 1.2 \text{ deg} = 0.0209 \text{ rad}. \tag{10.6.2}$$

The model parameters have the following values:

$$\begin{aligned} K_t &= 0.19 \text{ Nm/A}, \\ K_v &= 0.22 \text{ Vs/rad}, \\ R &= 2.00 \text{ ohms}, \\ L &= 1.40 \text{ mH}, \\ J &= 0.7 \times 10^{-3} \text{ kgm}^2, \\ N &= 7, \end{aligned} \tag{10.6.3}$$

so the plant transfer function $G(s)$ is

$$G(s) = \frac{27697}{s(s^2 + 1429s + 42653)}. \tag{10.6.4}$$

The plant transfer function $G(s)$ in combination with the zero-order hold unit may be represented by the discrete transfer function

$$G(z) = \mathcal{Z}\left[\frac{1 - \exp(Ts)}{s} G(s)\right]. \tag{10.6.5}$$

From (10.5.7) and (10.6.2)

$$T \leq 0.0209 \text{ s}. \tag{10.6.6}$$

Thus the design was initially tried with a sampling rate $T = 20$ ms. This was found to give unsatisfactory results; although (10.5.3) was satisfied, the MBP was unable to satisfy (10.5.1). Hence the design was repeated with $T = 10$ ms and $T = 5$ ms, both of which gave acceptable results with (10.5.1) satisfied. For $T = 5$ ms, this was achieved with the initial design. The results are shown in Table 10.1. The step responses $e(0, t, h)$ with the sampling points of the final designs are shown in Figures 10.5 to 10.7. Plots of $\tilde{e}(t_s)$ against t_s of the final designs are shown in Figures 10.8 to 10.10.

10.7 Concluding Remarks

The performance measure ϕ_e is a natural indicator of a control systems performance. This chapter extends the results of Zakian [Zak86a] to certain sampled-data systems, to provide a means of calculating ϕ_e. The design example shows that it is important to consider the performance over all time, particularly when the sampling time is comparatively long. As would be expected, the difference between ϕ_e and ϕ_e^* decreases with decreasing sampling period, as the system controller becomes closer to a continuous time controller.

Morari and Zafiriou [MZ89] detail how sampled-data control systems can be designed using H^∞ and H_2 techniques which consider performance over all time. Additional analysis of intersample behaviour using H^∞, H_2 and L^1 techniques can be found in [BPFT91], [LPF91], [DF92] and [Lin92].

		$T = 20$ ms	$T = 10$ ms	$T = 5$ ms
Design Stage (i)	$C(z)$	$325\frac{z^2-0.064z-0.260}{z^2+1.415z+0.448}$	$703\frac{z^2+0.136z-0.498}{z^2+1.100z+0.436}$	$600\frac{z^2-0.642z-0.150}{z^2+0.330z-0.068}$
	ϕ_e^*	0.0206 rad	0.0209 rad	0.0205 rad
	ϕ_e	0.1751 rad	0.0212 rad	0.0205 rad
Design Stage (ii)	$C(z)$	$177\frac{z^2+0.295z-0.455}{z^2+1.387z+0.461}$	$703\frac{z^2+0.136z-0.498}{z^2+1.210z+0.436}$	as above
	ϕ_e	0.0297 rad	0.0201 rad	0.0205 rad

Table 10.1: Design results

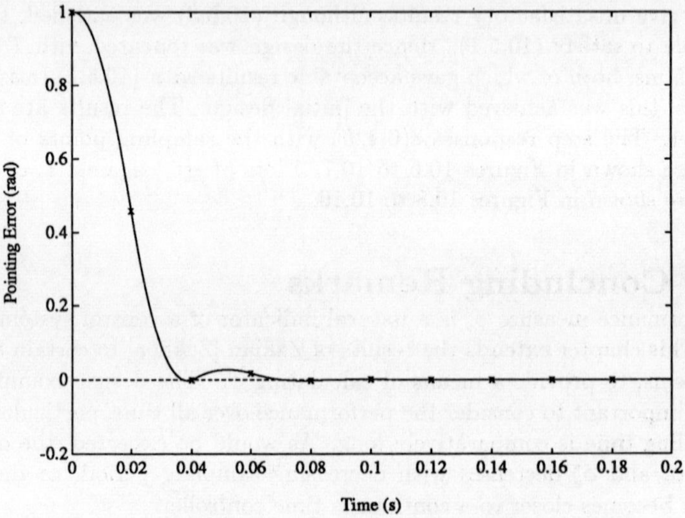

Figure 10.5: Step response of design showing sampling points for $T = 20$ ms

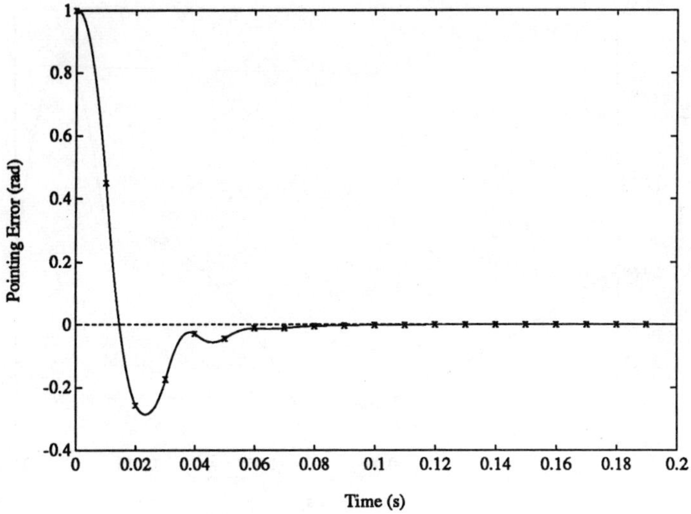

Figure 10.6: Step response of design showing sampling points for $T = 10$ ms

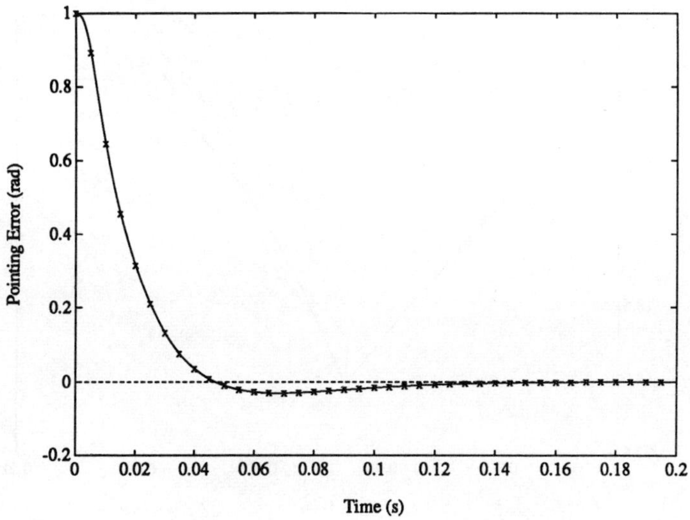

Figure 10.7: Step response of design showing sampling points for $T = 5$ ms

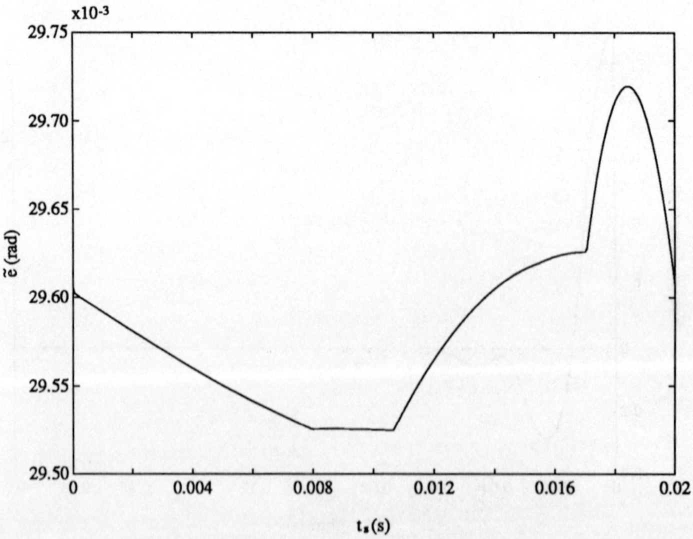

Figure 10.8: $\tilde{e}(t_s)$ against t_s for $T = 20$ ms

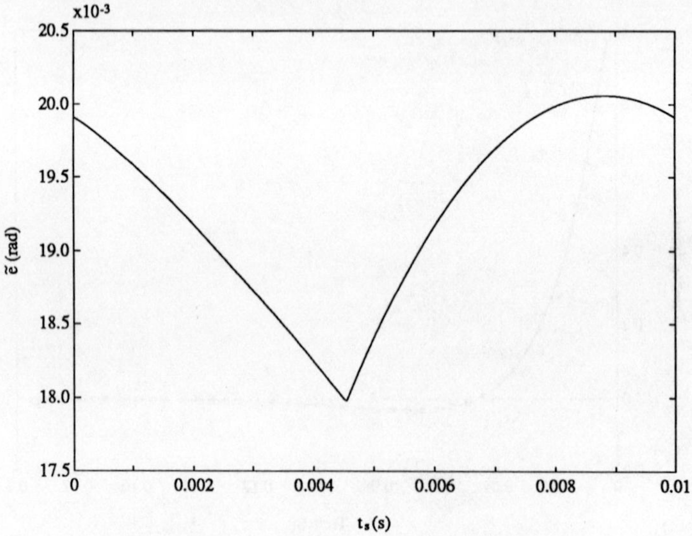

Figure 10.9: $\tilde{e}(t_s)$ against t_s for $T = 10$ ms

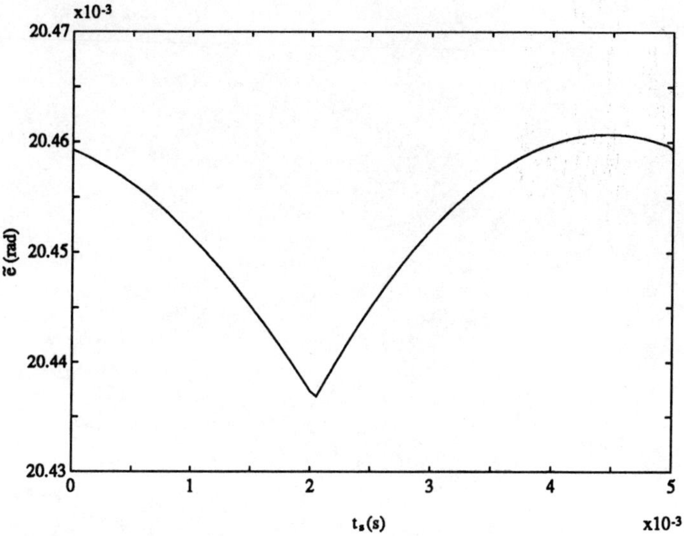

Figure 10.10: $\tilde{e}(t_s)$ against t_s for $T = 5$ ms

Chapter 11

Frequency Domain Approaches for Continuous Time Systems

11.1 Introduction

In this chapter, the frequency domain design of continuous time critical control systems is dealt with. The time domain design of these systems using Zakian's framework is described in Chapter 8.

Recently, H^∞-optimisation, which was introduced by Zames [Zam81] and developed by Zames [ZF83], Francis [FZ84, Fra87, FHZ84], Vidyasagar [Vid85a], Kwakernaak [Kwa85], Grimble [Gri86], Glover [GD88], Doyle [DGKF89] and others, has been applied to the design of many continuous systems. It mainly deals with the design of a feedback controller that minimises the H^∞-norm of the transfer function matrix from an exogenous input to an output to be controlled, which represents many design problems, for example, the weighted sensitivity problem [Zam81], the mixed sensitivity problem [VJ84, Kwa85], the tracking problem [Vid85b, WP86], the robust stabilisation problem [Kim84], the model-matching problem [Fra87]. Therefore, this chapter combines the advantages of Zakian's framework and the H^∞-optimisation method. Frequency domain design approaches for critical systems are developed [Liu90, Liu91, Liu93]. Some relations are established between an input space, a controller and output performance. These relations show that the problem of designing critical control systems with one design criterion can be converted into a standard H^∞-synthesis problem.

Two input spaces in the frequency domain were defined in Section 8.2.7. The relationship between the time domain and the frequency domain under some conditions is given in Section 11.2. The formulae for evaluation of output performance of the systems are established in Section 11.3. Section 11.4 looks at the controller design. Finally, in Section 11.5 simulation examples are given to demonstrate these frequency domain approaches.

11.2 Relationship between the Time Domain and the Frequency Domain

In a critical control system, the generalised output is required to be bounded in the time domain. In order to design a critical system in the frequency domain, it is important to find a relationship between the time domain and the frequency domain. Before the relationship is given, the following lemmas are introduced.

Lemma 11.1 *Let h be a continuous function such that $|h(t)| \leq D$ for $t > 0$ and $h(t) = 0$ for $t \leq 0$, and $H \in \mathbb{H}^1$. Then*

$$h(t) = \frac{1}{2\pi} \int_{-\infty}^{+\infty} H(j\omega) \exp(j\omega t) \, d\omega. \tag{11.2.1}$$

Proof: If there is a continuous function h which satisfies that $|h(t)| \leq D \exp(rt)$ $(r \geq 0)$ for $t > 0$ and $h(t) = 0$ for $t \leq 0$, then there exist Laplace transform pairs [Doe70]:

$$H(s) = \int_{0}^{+\infty} h(t) \exp(-st) \, dt, \tag{11.2.2}$$

$$h(t) = \frac{1}{2\pi j} \int_{r-j\infty}^{r+j\infty} H(s) \exp(st) \, ds. \tag{11.2.3}$$

In (11.2.3), let $s = r + j\omega$, then

$$h(t) = \frac{1}{2\pi} \int_{-\infty}^{+\infty} H(r + j\omega) \exp(rt + j\omega t) \, d\omega. \tag{11.2.4}$$

Multiplying both sides of (11.2.4) by $\exp(-rt)$ gives

$$h(t) \exp(-rt) = \frac{1}{2\pi} \int_{-\infty}^{+\infty} H(r + j\omega) \exp(j\omega t) \, d\omega. \tag{11.2.5}$$

From $|h(t)| \leq D$, it is clear that $r = 0$. By $H \in \mathbb{H}^1$ and $r = 0$, the lemma is proved. □□

In Laplace transformation, the inverse Laplace transform of H is defined by (11.2.3). In Lemma 11.1, since it is assumed that $H \in \mathbb{H}^1$, the inverse Laplace transform of H can be simply expressed by (11.2.1).

Lemma 11.2 *If $P \in \mathbb{H}^n$ $(n \in [1, \infty])$, then $PQ \in \mathbb{H}^m$, $\forall Q \in \mathbb{H}^m$ $(m \in [1, \infty])$.*

Proof: By the definition of the Hardy space $\mathbb{H}^m$ in Section 2.7, then

$$\| PQ \|_m \leq \| P \|_\infty \| Q \|_m. \tag{11.2.6}$$

Since $P \in \mathbb{H}^n$ and $Q \in \mathbb{H}^m$,

$$\| P \|_\infty < \infty \quad \text{and} \quad \| Q \|_m < \infty. \tag{11.2.7}$$

Therefore,

$$\| PQ \|_m < \infty. \tag{11.2.8}$$

The lemma is proved. □□

In fact, the result of Lemma 11.2 is obvious from the definition of the Hardy space $\mathbb{H}^m$.

Lemma 11.3 *Suppose* $P \in \mathbb{H}^n$ $(n \in [1, \infty))$, *then*

$$\lim_{n\to\infty} \| P \|_n = \| P \|_\infty. \tag{11.2.9}$$

Proof: If $\| P \|_\infty = 0$, it is clear that (11.2.9) is true. Now, assume $\| P \|_\infty \neq 0$ and $< \infty$. Let $Q = P \| P \|_\infty^{-1}$, then $\| Q \|_\infty = 1$. For any $\varepsilon > 0$, there exist $\omega_0 \in \mathbb{R}$ and $\delta \in \mathbb{R}^+$ satisfying that $|Q| > 1 - \varepsilon$, for all $\omega \in (\omega_0 - \delta, \omega_0 + \delta)$.

Hence

$$\begin{aligned} \| Q \|_n &\geq \left[\frac{1}{2\pi} \int_{\omega_0-\delta}^{\omega_0+\delta} |Q(j\omega)|^n \, \mathrm{d}\omega \right]^{1/n} \\ &\geq \left[\frac{1}{\pi} \delta (1-\varepsilon)^n \right]^{1/n} \\ &= (1-\varepsilon) \left(\frac{\delta}{\pi} \right)^{1/n}. \end{aligned} \tag{11.2.10}$$

Since $\delta > 0$ and as $n \to \infty$, $(\delta\pi^{-1})^{1/n} \to 1$, then there exists N_0 satisfying that $(\delta\pi^{-1})^{1/n} > 1 - \varepsilon$, for $n \geq N_0$. Thus, for $n \geq N_0$, $\| Q \|_n > (1-\varepsilon)(1-\varepsilon) > 1 - 2\varepsilon$. By the above, it can be concluded that

$$\lim_{n\to\infty} \| Q \|_n \geq 1. \tag{11.2.11}$$

Since $\| Q \|_n < \infty$ and $|Q| \leq 1$, there exists a M_0 satisfying

$$\frac{1}{2\pi} \left(\int_{-\infty}^{-M_0} |Q(j\omega)|^n \, \mathrm{d}\omega + \int_{M_0}^{+\infty} |Q(j\omega)|^n \, \mathrm{d}\omega \right) \leq 1. \tag{11.2.12}$$

Hence

$$\begin{aligned} \| Q \|_n &\leq \left[\frac{1}{2\pi} \int_{-M_0}^{M_0} |Q(j\omega)|^n \, \mathrm{d}\omega + 1 \right]^{1/n} \\ &\leq \left(\frac{M_0}{\pi} + 1 \right)^{1/n} \end{aligned} \tag{11.2.13}$$

As $n \to \infty$, $(M_0\pi^{-1}+1)^{1/n} \to 1$. Then

$$\lim_{n\to\infty} \| Q \|_n \leq 1. \tag{11.2.14}$$

According to (11.2.11) and (11.2.14), it follows that

$$\lim_{n\to\infty} \| Q \|_n = 1. \tag{11.2.15}$$

Since $\| P \|_n = \| P \|_\infty \| Q \|_n$, with (11.2.15), it can be obtained that

$$\lim_{n\to\infty} \| P \|_n = \| P \|_\infty. \tag{11.2.16}$$

□□

In Lebesgue space L^n ($n \in [1,\infty]$),

$$\lim_{n\to\infty} \| f \|_n \tag{11.2.17}$$

exists if and only if f is equivalent to a bounded function [Eps70]. So, the result of Lemma 11.3 is obvious.

Theorem 11.1 *Let $v = \mathcal{L}^{-1}[PQ]$ and suppose that $P \in \mathbb{H}^\infty$ and $Q \in \mathbb{H}^1(D)$. Then*

$$\sup\left\{|v(t)| : t \in \mathbb{R}^+, Q \in \mathbb{H}^1(D)\right\} \leq \| P \|_\infty D. \tag{11.2.18}$$

Proof: It follows from Lemmas 1 and 2 that $PQ \in \mathbb{H}^1$ and

$$v(t) = \frac{1}{2\pi}\int_{-\infty}^{+\infty} P(j\omega)Q(j\omega)\exp(j\omega t)\, d\omega. \tag{11.2.19}$$

Thus

$$\begin{aligned} |v(t)| &\leq \frac{1}{2\pi}\int_{-\infty}^{+\infty} |P(j\omega)||Q(j\omega)|\, d\omega \\ &\leq \frac{1}{2\pi}\sup\{|P(j\omega)| : \omega \in \mathbb{R}\}\int_{-\infty}^{+\infty} |Q(j\omega)|\, d\omega \\ &= \| P \|_\infty \| Q \|_1. \end{aligned} \tag{11.2.20}$$

Therefore, for $Q \in \mathbb{H}^1(D)$,

$$\sup\left\{|v(t)| : t \in \mathbb{R}^+, Q \in \mathbb{H}^1(D)\right\} \leq \| P \|_\infty D. \tag{11.2.21}$$

□□

Theorem 11.2 *Let $v = \mathcal{L}^{-1}[PQ]$ and suppose that $P \in \mathbb{H}^l$ ($l \in [1,\infty)$) and $Q \in \mathbb{H}^1(D)$. Then*

$$\sup\left\{|v(t)| : t \in \mathbb{R}^+, Q \in \mathbb{H}^1(D)\right\} = \| P \|_\infty D. \tag{11.2.22}$$

Proof: It follows from Lemmas 1 and 2 that $PQ \in \mathbb{H}^1$ and

$$v(t) = \frac{1}{2\pi}\int_{-\infty}^{+\infty} P(j\omega)Q(j\omega)\exp(j\omega t)\, d\omega. \tag{11.2.23}$$

Thus

$$|v(t)| \leq \frac{1}{2\pi}\int_{-\infty}^{+\infty} |P(j\omega)|\,|Q(j\omega)|\, d\omega. \tag{11.2.24}$$

An application of Holder's inequality to (11.2.24) gives

$$|v(t)| \leq \| P \|_n \| Q \|_m \tag{11.2.25}$$

where $n \geq l$ and $m = n(n-1)^{-1}$.

Assume that $\| Q \|_m \leq D_m < \infty$ and $D_m \to D$, as $m \to 1$. Then

$$\sup\left\{|v(t)| : t \in \mathbb{R}^+, Q \in \mathbb{H}^m(D_m)\right\} \leq \| P \|_n D_m. \tag{11.2.26}$$

For every $t \in \mathbb{R}^+$, there is a function $Q_t(j\omega)$ defined by

$$Q_t(j\omega) = \frac{|P(j\omega)|^{n-2}\, P(-j\omega)\exp(-j\omega t)}{\| P \|_n^{n-1}} D_m. \tag{11.2.27}$$

It is clear that $\| Q_t \|_m = D_m$, i.e. $Q_t \in \mathbb{H}^m(D_m)$. Equations (11.2.23) and (11.2.27) give

$$|v(t)| = \| P \|_n D_m. \tag{11.2.28}$$

Thus

$$\sup\left\{|v(t)| : t \in \mathbb{R}^+, Q \in \mathbb{H}^m(D_m)\right\} \geq \| P \|_n D_m. \tag{11.2.29}$$

From (11.2.26) and (11.2.29), it follows that

$$\sup\left\{|v(t)| : t \in \mathbb{R}^+, Q \in \mathbb{H}^m(D_m)\right\} = \| P \|_n D_m. \tag{11.2.30}$$

As $n \to \infty$, then

$$\lim_{n\to\infty} \| Q \|_{n(n-1)^{-1}} = \| Q \|_1 \leq D. \tag{11.2.31}$$

By Lemma 11.3, it can be concluded that

$$\sup\left\{|v(t)| : t \in \mathbb{R}^+, Q \in \mathbb{H}^1(D)\right\} = \| P \|_\infty D. \tag{11.2.32}$$

□□

Theorems 11.1 and 11.2 establish the relationships between the time domain and the frequency domain.

11.3 Evaluation of Output Performance

The standard feedback system $S_C(G,K)$ is shown in Figure 2.2. The system $S_C(G,K)$ with only the reference input R is considered first. The system $S_C(G,K)$ with both reference input R and disturbance Ω is then discussed.

Following Zakian [Zak86a], in continuous time systems, the measure of performance of the system $S_C(G,K)$ with external inputs in $\mathcal{F}$ is defined as

$$\phi_C(K) = \sup\left\{\left|v(t,\bar{W},K)\right| : t \in \mathbb{R}, \bar{W} \in \mathcal{F}\right\}. \tag{11.3.1}$$

where the output $v : \mathbb{R} \to \mathbb{R}$ depends on the external inputs $\bar{W}$ and the controller K.

From the above, clearly, $\phi_C(K)$ is the largest absolute value of the generalised output v for all $\bar{W} \in \mathcal{F}$ over all time $t \in \mathbb{R}$.

Here, it is assumed that there is only the reference input R. The generalised output V is replaced by the error E. Thus, in Figure 2.2, the input-output rule $\mathbf{R}$ is expressed by the transfer function $(1+GK)^{-1}$ which relates the input R and the error E:

$$E(R,K) = (1+GK)^{-1}R. \tag{11.3.2}$$

In critical systems, the main objective is to ensure that during all time t, the absolute error $|e(t,R,K)|$ remains tolerably small. A measure of performance of the feedback system $S_C(G,K)$ as given by (11.3.1) becomes

$$\phi_C(K) = \sup\left\{|e(t,R,K)| : t \in \mathbb{R}^+, R \in \mathcal{F}\right\}, \tag{11.3.3}$$

where $\mathcal{F}$ is an input space in the frequency domain.

The evaluation of the number $\phi_C(K)$ is difficult if effected directly from the defining expression (11.3.3). This is mainly because of the supremum operation required over the input function space $\mathcal{F}$. Several computationally simple expressions for $\phi_C(K)$ in the time domain are given in Section 8.5. In this section, some formulae for the evaluation of $\phi_C(K)$ in the frequency domain are presented. The input space plays an important part in a critical system. Two input spaces $\mathcal{F}(F,D)$ and $\mathcal{F}(D_1,D_2)$ in the frequency domain, which are defined in Section 8.2.7, are considered here. We first consider the system $S_C(G,K)$ with only the reference input $R \in \mathcal{F}(F,D)$.

Theorem 11.3 *Let input-output rule* $\mathbf{R} : R \mapsto E(R,K)$ *be given by (11.3.2) for all* $R \in \mathcal{F}(F,D)$ *and suppose* $(1+GK)^{-1}F \in \mathbb{H}^n$ $(n \in [1,\infty))$. *Then*

$$\phi_C(K) = \left\|(1+GK)^{-1}F\right\|_\infty D. \tag{11.3.4}$$

Proof: By Definition 8.6, $R = FW_1$, for $W_1 \in \mathbb{H}^1(D)$ and given $F \in \mathbb{R}(s)$. Then, it follows from (11.3.1) that

$$e = \mathcal{L}^{-1}\left[(1+GK)^{-1}FW_1\right]. \tag{11.3.5}$$

Let $P = (1+GK)^{-1}F$ and $Q = W_1$. Then, by Theorem 11.2,

$$\sup\left\{|e(t,R,K)| : t \in \mathbb{R}^+, R \in \mathcal{F}(F,D)\right\} = \left\| (1+GK)^{-1}F \right\|_\infty D. \qquad (11.3.6)$$

which, with (11.3.3), proves the theorem. □□

Theorem 11.4 *Let input-output rule* $\mathbf{R} : R \mapsto E(R,K)$ *be given by (11.3.2) for all* $R \in \mathcal{F}(F,D)$ *and suppose* $(1+GK)^{-1}F \in \mathbb{H}^\infty$. *Then*

$$\phi_C(K) \leq \left\| (1+GK)^{-1}F \right\|_\infty D. \qquad (11.3.7)$$

Proof: This is similar to Theorem 11.3. The difference is that Theorem 11.1 is used instead of Theorem 11.2. □□

Theorems 11.3 and 11.4 show that if the system $S_C(G,K)$ is subjected to reference inputs in the input space $\mathcal{F}(F,D)$, the computation of $\phi_C(K)$ in the time domain can be converted to computing $\| (1+GK)^{-1}F \|_\infty$ in the frequency domain.

The evaluation of output performance for the system $S_C(G,K)$ with only reference inputs in the input space $\mathcal{F}(D_1,D_2)$ is now considered; however, the following lemmas are required.

Lemma 11.4 *If* $Q \in \mathbb{H}^2$, *then* $\| Q \|_2 = \| q \|_2$, *i.e.*

$$\left[\frac{1}{2\pi}\int_{-\infty}^{+\infty} |Q(j\omega)|^2 \,\mathrm{d}\omega\right]^{1/2} = \left[\int_0^{+\infty} q^2(t)\,\mathrm{d}t\right]^{1/2}. \qquad (11.3.8)$$

Proof: Refer to Plancherel's theorem. □□

Lemma 11.5 *If* $P \in \mathbb{H}^\infty$, *then*

$$\| P \|_\infty = \sup\left\{\| PQ \|_2 : Q \in \mathbb{H}^2(D)\right\}/D. \qquad (11.3.9)$$

Proof: If $P \in \mathbb{H}^\infty$, then [Fra87]

$$\| P \|_\infty = \sup\left\{\| PX \|_2 : X \in \mathbb{H}^2, \| X \|_2 \leq 1\right\}. \qquad (11.3.10)$$

Since $Q \in \mathbb{H}^2(D)$, $\| Q/D \|_2 \leq 1$. Hence,

$$\begin{aligned}\| P \|_\infty &= \sup\left\{\| PQ/D \|_2 : Q/D \in \mathbb{H}^2, \| Q/D \|_2 \leq 1\right\} \\ &= \sup\left\{\| PQ \|_2/D : Q \in \mathbb{H}^2(D)\right\} \\ &= \sup\left\{\| PQ \|_2 : Q \in \mathbb{H}^2(D)\right\}/D. \qquad (11.3.11)\end{aligned}$$

□□

A relationship between the input space $\mathcal{F}(D_1, D_2)$, the controller K and the output performance $\phi_C(K)$ is given by the following theorem.

Theorem 11.5 *Assume that* $(1+GK)^{-1} \in \mathbb{R}\mathbb{H}^\infty$ *and* $R \in \mathcal{F}(D_1, D_2)$, *then*

$$\phi_C(K) = \left\| (1+GK)^{-1} \right\|_\infty D_1 D_2. \tag{11.3.12}$$

Proof: By Definition 8.7, $R = F_1 F_2$, for $F_1 \in \mathbb{H}^2(D_1)$ and $F_2 \in \mathbb{H}^2(D_2)$. Let $H = (1+GK)^{-1}F_1$. By Lemma 11.2, then $H \in \mathbb{H}^2$. By the convolution, then the error of the system is

$$e(t, R, K) = \int_0^t h(t-\tau) f_2(\tau)\, \mathrm{d}\tau. \tag{11.3.13}$$

Using the Cauchy-Schwarz inequality [Lue69] gives

$$\begin{aligned} |e(t,R,K)| &\le \left[\int_0^t h^2(\tau)\,\mathrm{d}\tau\right]^{1/2} \left[\int_0^t f_2^2(\tau)\,\mathrm{d}\tau\right]^{1/2} \\ &\le \left[\int_0^\infty h^2(\tau)\,\mathrm{d}\tau\right]^{1/2} \left[\int_0^\infty f_2^2(\tau)\,\mathrm{d}\tau\right]^{1/2} \\ &= \| h \|_2 \| f_2 \|_2. \end{aligned} \tag{11.3.14}$$

By Lemma 11.4, then

$$|e(t,R,K)| \le \left\| (1+GK)^{-1} F_1 \right\|_2 \| F_2 \|_2. \tag{11.3.15}$$

Since $R \in \mathcal{F}(D_1, D_2)$,

$$|e(t,R,K)| \le \left\| (1+GK)^{-1} F_1 \right\|_2 D_2. \tag{11.3.16}$$

According to the definition of the performance function $\phi_C(K)$, it can be deduced from (11.3.3) and (11.3.16) that

$$\phi_C(K) \le \gamma D_2, \tag{11.3.17}$$

where

$$\gamma = \sup\left\{ \left\| (1+GK)^{-1} F_1 \right\|_2 : F_1 \in \mathbb{H}^2(D_1) \right\}. \tag{11.3.18}$$

For $\sigma \in (0, \gamma)$ and $\alpha > 1$, let

$$H_n^+ = (1+GK)^{-1} F_{1n}^+, \quad n \in \mathbb{N}^+, \tag{11.3.19}$$

where $F_{1n}^+ \in \mathbb{H}^2(D_1)$ and makes H_n^+ satisfy

$$\left\| H_n^+ \right\|_2 = \gamma - \sigma\alpha^{-n}. \tag{11.3.20}$$

Since γ is the supremum of $\| H \|_2$ for $F_1 \in \mathbb{H}^2(D_1)$, according to the definition of the supremum there certainly exists a sequence $F_{1n}^+ \in \mathbb{H}^2(D_1)$ $(n \in \mathbb{N}^+)$ which makes (11.3.20) hold.

Choose a particular input sequence $R_n^* = F_{1n}^* F_{2n}^*$ $(n \in \mathbb{N}^+)$ defined by $F_{1n}^* = F_{1n}^+$ and $F_{2n}^* = \mathcal{L}\left[f_{2n}^+\right]$, where

$$f_{2n}^+(\tau) = \begin{cases} h_n^+(t-\tau) D_2 \| H_n^+ \|_2^{-1} & \text{for} \quad 0 \le \tau \le t \\ 0 & \text{for} \quad \tau > t \quad \text{and} \quad \tau < 0 \end{cases} \tag{11.3.21}$$

and $h_n^+ = \mathcal{L}^{-1}[H_n^+]$. It is obvious that $F_{2n}^+ \in \mathbb{H}^2(D_2)$. So, $R_n^* \in \mathcal{F}(D_1, D_2)$ for $n \in \mathbb{N}^+$.

Substituting (11.3.21) into (11.3.13) gives

$$e(t, R_n^*, K) = \int_0^t D_2 (h_n^+(\tau))^2 \left\| H_n^+ \right\|_2^{-1} \mathrm{d}\tau. \tag{11.3.22}$$

As $t \to \infty$, then

$$\begin{aligned} e(\infty, R_n^*, K) &= \int_0^\infty D_2 (h_n^+(\tau))^2 \left\| H_n^+ \right\|_2^{-1} \mathrm{d}\tau \\ &= D_2 \left\| h_n^+ \right\|_2^2 \left\| H_n^+ \right\|_2^{-1}. \end{aligned} \tag{11.3.23}$$

By Lemma 11.4, then

$$\begin{aligned} e(\infty, R_n^*, K) &= D_2 \left\| H_n^+ \right\|_2 \\ &= \left(\gamma - \sigma\alpha^{-n}\right) D_2. \end{aligned} \tag{11.3.24}$$

As $n \to \infty$, then

$$e(\infty, R_\infty^*, K) = \gamma D_2. \tag{11.3.25}$$

With (11.3.3), it follows from the above that

$$\phi_C(K) \ge \gamma D_2. \tag{11.3.26}$$

From (11.3.17) and (11.3.26), it can be concluded that

$$\phi_C(K) = \sup\left\{\left\| (1+GK)^{-1} F_1 \right\|_2 : F_1 \in \mathbb{H}^2(D_1)\right\} D_2. \tag{11.3.27}$$

Using Lemma 11.5 gives

$$\phi_C(K) = \left\| (1+GK)^{-1} \right\|_\infty D_1 D_2. \tag{11.3.28}$$

Therefore, the theorem is established. □□

Theorem 11.5 shows that the design of the controller K that minimises $\phi_C(K)$ for the system $S_C(G,K)$, subjected to the input space $\mathcal{F}(D_1,D_2)$, is converted to the problem of minimising $\|(1+GK)^{-1}\|_\infty$.

Now, it is assumed that there are both reference input R and disturbance Ω in the system $S_C(G,K)$. Let the external input vector $\bar{W}=[R,\Omega]$. Then, the input-output rule $\mathbf{R}:\bar{W}\mapsto E(\bar{W},K)$ is expressed as

$$E(\bar{W},K)=(1+GK)^{-1}(R-\Omega). \tag{11.3.29}$$

The measure of output performance (11.3.1) is rewritten as

$$\phi_{CC}(K)=\sup\left\{|e(t,R,K)|:t\in\mathbb{R}^+,\bar{W}\in\mathcal{F}\right\}, \tag{11.3.30}$$

where $\mathcal{F}$ denotes an input space in the frequency domain.

In this case, we have the following theorem:

Theorem 11.6 *For the system $S_C(G,K)$ with reference input R and disturbance Ω, then*

$$\phi_{CC}(K)=\left\|(1+GK)^{-1}F\right\|_\infty(D_1+D_2),\quad \text{for}\quad \mathcal{F}=\mathcal{F}(F,D_1)\times\mathcal{F}(F,D_2) \tag{11.3.31}$$

and

$$\phi_{CC}(K)=\left\|(1+GK)^{-1}\right\|_\infty(D_1D_2+D_3D_4),\quad \text{for}\quad \mathcal{F}=\mathcal{F}(D_1,D_2)\times\mathcal{F}(D_3,D_4). \tag{11.3.32}$$

Proof: From (11.3.29), we have

$$E=(1+GK)^{-1}R-(1+GK)^{-1}\Omega. \tag{11.3.33}$$

It can be seen from the above that the error E consists of two parts: $(1+GK)^{-1}R$ and $(1+GK)^{-1}\Omega$. The former is the effect of the reference input R and the latter is the effect of the disturbance Ω. Applying the results of Theorems 11.4 and 11.5 to the two parts above, respectively, then the theorem follows immediately. □□

The evaluation formulae of output performance given by the theorems in this section display that the design of the optimal controller K which minimises the performance function $\phi_C(K)$ or $\phi_{CC}(K)$ is transferred to the problem of minimising $\|(1+GK)^{-1}F\|_\infty$ or $\|(1+GK)^{-1}\|_\infty$ by the H^∞-optimisation method. For discussions of the H^∞-optimisation method, see, for example, Zames [Zam81], Francis [FZ84, Fra87], Vidyasagar [Vid85a] and Doyle [DGKF89].

11.4 Controller Design

Suppose $G \in \mathbb{R}(s)$, so that G is the transfer function of a scalar, linear, time-invariant system. Suppose that a controller $K \in \mathbb{R}(s)$ is connected to the plant G in the feedback configuration shown in Figure 2.2.

The plant can naturally be modelled as a quotient of coprime polynomials [Ros70], $G(s) = A(s)/B(s)$. Since our ultimate goal is a stable system, the plant also can be expressed as

$$G(s) = \frac{B_n(s)}{B_d(s)}, \tag{11.4.1}$$

$$B_n(s) = \frac{A(s)}{M(s)}, \tag{11.4.2}$$

$$B_d(s) = \frac{B(s)}{M(s)}, \tag{11.4.3}$$

where $M(s)$ is a strictly Hurwitz polynomial of degree equal to the degrees of $B(s)$. Moreover, since $A(s)$ and $B(s)$ are coprime, the rational functions $B_n, B_d \in \mathbb{R}\mathbb{H}^\infty$ are coprime in the sense that there exist $C_n, C_d \in \mathbb{R}\mathbb{H}^\infty$ such that

$$C_n B_n + C_d B_d = 1. \tag{11.4.4}$$

Therefore, the set of all controllers that stabilise G, denoted by $\mathbb{F}(Q)$, is given by [YJB76, Vid85a]

$$\mathbb{F}(Q) = \left\{ K = \frac{C_n + QB_d}{C_d - QB_n} : Q \in \mathbb{R}\mathbb{H}^\infty \quad \text{and} \quad C_d - QB_n \neq 0 \right\}. \tag{11.4.5}$$

First, assume that the reference input $R \in \mathcal{F}(F, D)$ in the system $S_C(G, K)$ without the disturbance Ω. By Theorems 11.3 and 11.4, then the objective is to find a $K \in \mathbb{F}(Q)$ that minimises the following function

$$\beta = \left\| (1 + GK)^{-1} F \right\|_\infty. \tag{11.4.6}$$

Substituting (11.4.6) by (11.4.1), (11.4.4) and (11.4.5) gives

$$\beta = \| (C_d - QB_n) B_d F \|_\infty. \tag{11.4.7}$$

Let $Q_1 = C_d B_d F$ and $Q_2 = B_n B_d F$.

Bring in an inner-outer factorisation [SNF70] of $Q_2 : Q_2 = Q_{2i} Q_{2o}$, where Q_{2i} is inner, i.e. $Q_{2i}(-s) Q_{2i}(s) = 1$, and Q_{2o} is outer, i.e. it has no zero in Re $s \leq 0$. For $Q \in \mathbb{R}\mathbb{H}^\infty$, we have

$$\beta = \| H - Q_{2o} Q \|_\infty, \tag{11.4.8}$$

where $H := Q_{2i}^{-1} Q_1$.

Factorise H as $H = H_1 + H_2$ with H_1 strictly proper and analytic in Re $s \le 0$ and H_2 in $\mathbb{R}\mathbb{H}^\infty$. Then H_1 has a minimal state-space realisation

$$H_1 = \bar{C}(sI - \bar{A})^{-1}\bar{B} \tag{11.4.9}$$

with $\bar{A}$ antistable. The controllability and observability grammian matrices G_c and G_o are the unique solution of

$$\bar{A}G_c + G_c\bar{A}^T = \bar{B}\bar{B}^T, \tag{11.4.10}$$

$$\bar{A}^T G_o + G_o\bar{A} + \bar{C}^T\bar{C}. \tag{11.4.11}$$

Let λ^2 equal the largest eigenvalue of G_cG_o and let ξ be a corresponding eigenvector:

$$G_cG_o\xi = \lambda^2\xi. \tag{11.4.12}$$

Define the real-rational functions:

$$C_1 = \bar{C}(sI - \bar{A})^{-1}\xi, \tag{11.4.13}$$

$$C_2 = \bar{B}(sI - \bar{A})^{-1}G_o\xi/\lambda. \tag{11.4.14}$$

By Theorem 6.1 [Fra87], it can be obtained that

$$\inf\{\beta\} = \lambda. \tag{11.4.15}$$

with

$$Q = Q_{2o}(H - \lambda C_1/C_2). \tag{11.4.16}$$

Summarising the above procedure gives the following theorem.

Theorem 11.7 *Assume plant $G \in \mathbb{R}(s)$ and the reference input $R \in \mathcal{F}(F, D)$. Then there exists an optimal controller $K_o \in \mathbb{R}(s)$ which is of the form*

$$K_o(s) = \frac{C_n + Q_{2o}(H - \lambda C_1/C_2)B_d}{C_d - Q_{2o}(H - \lambda C_1/C_2)B_n}. \tag{11.4.17}$$

And

$$\phi_C(K_o) \le \lambda D. \tag{11.4.18}$$

Note that for the input space $\mathcal{F} = \mathcal{F}(F, D_1) \times \mathcal{F}(F, D_2)$, the minimisation of the performance function $\phi_{CC}(K)$ is same as the above. Similarly, if the input space $\mathcal{F} = \mathcal{F}(D_1, D_2)$ or $\mathcal{F} = \mathcal{F}(D_1, D_2) \times \mathcal{F}(D_3, D_4)$, we can also use the above procedure to design the optimal controller under setting $F = 1$.

Though only one H^∞-optimisation method which is mentioned in [Fra87] is introduced to design the optimal controller K_o, many other H^∞-optimisation methods can also be applied to the design of the optimal controller K_o. We do not intend to introduce them here.

11.5 Examples

The following simulation illustrates the calculation of $\phi_C(K)$ and $K_o(s)$, and the properties of the system by the method presented in this chapter. The plant G and the input space $\mathcal{F}(F, D)$ are taken to be

$$G(s) = \frac{(s-2)(s-3)}{s^2+2s+2}, \tag{11.5.1}$$

$$R \in \mathcal{F}\left(\frac{s+1}{s+0.1}, 0.05\right). \tag{11.5.2}$$

A coprime factorisation of $G(s)$ is that $B_n = G(s)$ and $B_d = 1$. From (11.4.4), it can be obtained that $C_n = 0$ and $C_d = 1$. Then,

$$Q_{2i} = \frac{(s-2)(s-3)}{(s+2)(s+3)}, \tag{11.5.3}$$

$$Q_{2o} = \frac{(s+1)(s+2)(s+3)}{(s+0.1)(s^2+2s+2)}, \tag{11.5.4}$$

$$H = \frac{s^3+6s^2+11s+6}{(s-2)(s-3)(s+0.1)}. \tag{11.5.5}$$

The minimal realisation of the antistable part of H is as follows:

$$\bar{A} = \begin{bmatrix} 2 & 0 \\ 0 & 3 \end{bmatrix}, \tag{11.5.6}$$

$$\bar{B} = \begin{bmatrix} -211.671429 \\ 311.709676 \end{bmatrix}, \tag{11.5.7}$$

$$\bar{C} = [1 \quad 1]. \tag{11.5.8}$$

Solving (11.4.10) and (11.4.11), the grammian matrices G_c and G_o are

$$G_c = \begin{bmatrix} 204.06164 & -221.19816 \\ -221.19816 & 249.73986 \end{bmatrix}, \tag{11.5.9}$$

$$G_0 = \begin{bmatrix} 0.25 & 0.20 \\ 0.20 & 0.17 \end{bmatrix}. \tag{11.5.10}$$

Then, the maximum eigenvalue λ^2 of $G_c G_o$ and a corresponding eigenvector ξ are $\lambda^2 = 3.0506547$ and $\xi = [1 \quad -0.9443426]^T$.

From (11.4.13), (11.4.14) and (11.4.16), it can be deduced that

$$C_1 = \frac{0.0556573s - 1.1113147}{(s-2)(s-3)}, \tag{11.5.11}$$

$$C_2 = \frac{-0.0556561s - 1.1113123}{(s+2)(s+3)}, \tag{11.5.12}$$

$$Q = 2.7466495\frac{s^2+2s+2}{(s+1)(s+19.967484)}. \tag{11.5.13}$$

Therefore, the controller is designed as

$$K_o(s) = \frac{0.7875501s^2 + 1.5751002s + 1.5751002}{-0.5008189s^2 + 9.9497825s + 1}. \tag{11.5.14}$$

According to Theorem 11.7, it can be concluded that the performance of the system is

$$\phi_C(K_o) \leq 0.08733061. \tag{11.5.15}$$

In the above, choose

$$W_1(s) = \frac{1}{100(s+0.1)(s+0.1)}. \tag{11.5.16}$$

We can calculate that $\| W_1 \|_1 = 0.05$. The simulation results are shown in Figures 11.1 and 11.2, where $r(t), e(t), y(t)$ and $u(t)$ are the reference input, the error, the output and the control of the system, respectively. From Figure 11.1, it is found that if the input is $r(t)$, the largest value of the error in the system is 0.06416230, which is less than 0.08733061.

From the above, we can see that the simulation results coincide with the theoretical results. So, the simulation results demonstrate the frequency domain approaches for critical control systems presented in this chapter.

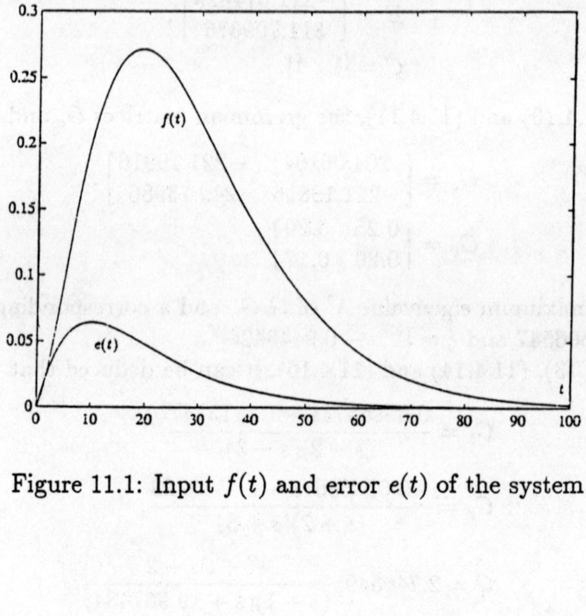

Figure 11.1: Input $f(t)$ and error $e(t)$ of the system

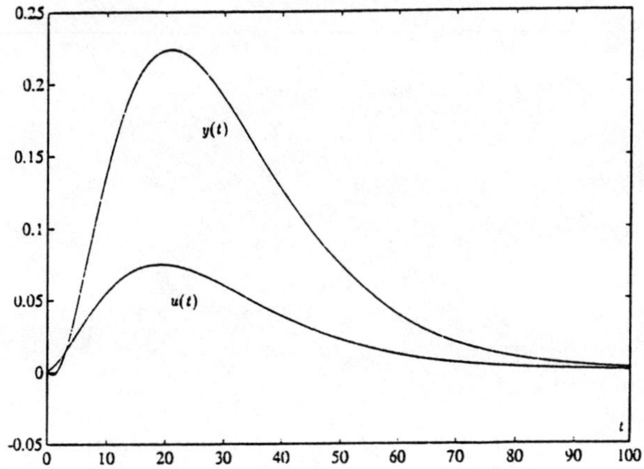

Figure 11.2: Output $y(t)$ and control $u(t)$ of the system

Bibliography

[AM71] B. D. O. Anderson and J. B. Moore. *Linear Optimal Control.* Prentice-Hall, New York, 1971.

[And69] N.I. Andreyev. *Correlation Theory of Statistically Optimal Systems.* W.B. Saunders Co., Philadelphia, PA., 1969.

[Arm84] D.S. Armstrong. Magnet/rail systems - a critical review of the options. In *IMechE Conf. Publ. 1984-12 - Proc. Int. Conf. on Maglev Transport Now and for the Future*, pages 59–66, Solihull, UK, 1984.

[Ast70] K.J. Astrom. *Introduction to Stochastic Control Theory.* Academic Press, New York, 1970.

[AW73] K.J. Astrom and B. Wittenmark. On self-tuning regulators. *Automatica*, 9:185–199, 1973.

[AW80] K.J. Astrom and B. Wittenmark. Self-tuning controllers based on pole-zero placement. *IEE Proc. D*, 127:120–130, 1980.

[Bad86] A.T. Bada. *Computer-Aided Design of Control Systems.* PhD thesis, University of Manchester Institute of Science and Technology, Manchester, 1986.

[Bad87a] A.T. Bada. CRITERIA : A package for computer-aided control systems design. *Computer-Aided Design*, 19(9):466–474, 1987.

[Bad87b] A.T. Bada. Robust brake control for a heavy-duty truck. *IEE Proc. D*, 134(1):1–8, 1987.

[BB91] S. Boyd and C. Barratt. *Linear Controller Design: Limits of Performance.* Prentice-Hall, Englewood Cliffs, N.J., 1991.

[BBN90] S. Boyd, C. Barratt, and S. Norman. Linear controller design: Limits of performance via convex optimization. *Proc. IEEE*, 78(3):573–578, 1990.

[BCS79] K.E. Bollinger, P.A. Cook, and P.S. Sandhu. Synchronous generator controller synthesis using moving boundary search techniques with frequency domain techniques. *IEEE Trans. Power Apparatus & Systems*, PAS-98(5):1497–1501, 1979.

[BHM79] R.G. Becker, A.J. Heunis, and D.Q. Mayne. Computer-aided design of control systems via optimization. *IEE Proc. D*, 126(6):573–578, 1979.

[BPFT91] B. Bamieh, J.B. Pearson, B.A. Francis, and A. Tannenbaum. A lifting technique for linear periodic systems with applications to sampled-data control. *Syst. Control Lett.*, 17:79–88, 1991.

[BR71] D.P. Bertsekas and I.B. Rhodes. On the minimax reachability of target sets and target tubes. *Automatica*, 7:233–247, 1971.

[CD82] T.R. Crossley and A.M.S. Dahshan. Design of a longitudinal ride-control systems by Zakian's method of inequalities. *J. Aircraft*, 19(9):730–738, 1982.

[CdWC90] C.C. Canudas de Wit and J. Carrillo. A modified EW-RLS algorithm for systems with bounded disturbances. *Automatica*, 26:599–606, 1990.

[CG75] D. W. Clarke and P.J. Gawthrop. Self-tuning controller. *Proc. IEE*, 122:929–934, 1975.

[CG79] D. W. Clarke and P.J. Gawthrop. Self-tuning control. *Proc. IEE*, 126:633–640, 1979.

[Chv83] V. Chvatal. *Linear Programming.* W. H. Freeman and Company, New York, 1983.

[CMT87] D.W. Clarke, C. Mohtadi, and P.S. Tuffs. Generalized predictive control — part I. The basic algorithm. *Automatica*, 22(2):137–148, 1987.

[Coe79] C.A.D. Coelho. Compensation of the speed governor of a water turbine by the method of inequalities. *ASME J. Dyn. Syst. Meas. & Control*, 101:205–211, 1979.

[Coh78] J.L. Cohon. *Multiobjective Programming and Planning*. Academic Press, New York, 1978.

[Dah81] A.M. Dahshan. A comparison between the method of inequalities and other techniques in control systems design. *Int. J. System Science*, 12(9):1149–1155, 1981.

[DF92] G.E. Dullerud and B.A. Francis. $\mathcal{L}_\infty$ analysis and design of sampled-data systems. *IEEE Trans. Automatic Control*, AC-37:436–446, 1992.

[DFT91] J.C. Doyle, B.A. Francis, and A.R. Tannenbaum. *Feedback Control Theory*. Macmillan, New York, 1991.

[DGKF89] J.C. Doyle, K. Glover, P.P. Khargonekar, and B.A. Francis. State space solutions to the standard H^2 and H^∞ control problems. *IEEE Trans. Automatic Control*, AC-34:831–847, 1989.

[Doe70] G. Doetsch. *Introduction to the Theory and Application of the Laplace Transformation*. Springer-Verlag, Heidelberg, 1970.

[DP87a] M.A. Dahleh and J.B. Pearson. l^1-optimal compensators for discrete-time systems. In *Proceedings of American Control Conference*, pages 1964–1968, Seattle, USA, 1987.

[DP87b] M.A. Dahleh and J.B. Pearson. l^1-optimal compensators for MIMO discrete-time systems. *IEEE Trans. Automatic Control*, AC-32:314–322, 1987.

[DP87c] M.A. Dahleh and J.B. Pearson. L^1-optimal compensators for continuous-time systems. *IEEE Trans. Automatic Control*, AC-32:889–895, 1987.

[Dur70] P.L. Duren. *Theory of H^p Spaces*. Academic Press, New York, 1970.

[Eps70] B. Epstein. *Linear Functional Analysis: An Introduction to Lebesgue Integration and Infinite-Dimensional Problems*. W. B. Saunders Co., 1970.

[FH82] E. Fogel and Y.F. Huang. On the value of information in system identification – bounded noise case. *Automatica*, 18(2):229–238, 1982.

[FHZ84] B. A. Francis, J. W. Helton, and G. Zames. H^∞-optimal feedback controllers for linear multivariable systems. *IEEE Trans. Automatic Control*, AC-29(10):888–900, 1984.

[Fle87] R. Fletcher. *Practical Methods of Optimization*. Wiley, Chichester, U.K., 2nd edition, 1987.

[FP80] G.F. Franklin and J.D. Powell. *Digital Control of Dynamic Systems*. Addison-Wesley, Reading, MA., 1980.

[FP86] P.J. Fleming and A.P. Pashkevich. Application of multi-objective optimization to compensator design for SISO control systems. *ElectronicsLetters*, 22(5):258–259, 1986.

[FPEN86] G.F. Franklin, J.D. Powell, and A. Emami-Naeini. *Feedback Control of Dynamic Systems*. Addison-Wesley, Reading, MA., 1986.

[Fra87] B.A. Francis. *A Course in H^∞ Control Theory*, volume 88 of *Lect. Notes Control & Inf. Sci.* Springer-Verlag, Berlin, 1987.

[Frö65] C.E. Fröberg. *Introduction to Numerical Analysis*. Addison-Wesley, Reading, MA., 1965.

[FZ84] B. A. Francis and G. Zames. On H^∞-optimal feedback theory for SISO feedback systems. *IEEE Trans. Automatic Control*, AC-29(1):9–16, 1984.

[GD88] K. Glover and J. C. Doyle. State space formula for all stabilizing controllers that satisfy an H^∞ norm bound and relations to risk sensitivity. *Syst. Control Lett.*, 11:167, 1988.

[Gem74] F.W. Gembicki. *Performance and Sensitivity Optimization: A Vector-Index Approach*. PhD thesis, Case Western Reserve University, Cleveland, Ohio, 1974.

[Gie78] D.P. Giesy. Calculation of Pareto-optimal solutions to multiple-objective problems using threshold-of-acceptability constraints. *IEEE Trans. Automatic Control*, AC-23(6):1114–1115, 1978.

[Gri86] M. J. Grimble. Optimal H^∞ robustness and the relationship to LQG design problems. *Int. J. Control*, 35(1):127–138, 1986.

[GS71] J. D. Glover and F. C. Schweppe. Control of linear dynamic systems with set constrained disturbances. *IEEE Trans. Automatic Control*, AC-16:411–422, 1971.

[GS84] G. C. Goodwin and K. S. Sin. *Adatpive Filtering Prediction and Control*. Prentice-Hall, Englewood Cliffs, N.J., 1984.

[Hen74] P. Henrici. *Applied and Computational Analysis*, volume 1. Wiley, New York, 1974.

[Hor63] I.M. Horowitz. *Synthesis of Feedback Systems*. Academic Press, London, 1963.

[Ino82] H. Inooka. Improved time-response algorithm for the method of inequalities. *Int. J. Control*, 35(1):127–138, 1982.

[Jay88] B.V. Jayawant. Are we in control? *IEE Proc. D*, 135(1):1–15, 1988.

[Kim84] H. Kimura. Robust stabilization for a class of transfer functions. *IEEE Trans. Automatic Control*, AC-29:788–783, 1984.

[Kip89] H. W. Kipp. *Narrow Fabric Weaving*. Arau, Frankfurt am Main, Sauerlander, Salzburg, 1989.

[KN82] G. Kreisselmeier and K. S. Narendra. Stable MRAC in the presence of bounded disturbances. *IEEE Trans. Automatic Control*, AC-27:1169–1175, 1982.

[KS79] G. Kreisselmeier and R. Steinhauser. Systematic control design by optimizing a vector performance index. In *Proc. IFAC Symp. on Computer-Aided Design of Control Systems*, pages 113–117, Zürich, Switzerland, 1979.

[KS83] G. Kreisselmeier and R. Steinhauser. Application of vector performance optimization to a robust control loop design for a fighter aircraft. *Int. J. Control*, 37(2):251–284, 1983.

[KU84] W. Kortüm and A. Utzt. Control law design and dynamic evaluations for a maglev vehicle with a combined lift and guidance suspension system. *ASME J. Dyn. Syst. Meas. & Control*, 106:286–292, 1984.

[Kwa85] H. Kwakernaak. Minimax frequency domain performance and robustness optimization of linear feedback systems. *IEEE Trans. Automatic Control*, AC-30(10):994–1004, 1985.

[Lan92] P.G. Lane. *Design of Control Systems with Input and Outputs Satisfying Certain Bounding Conditions*. PhD thesis, University of Manchester Institute of Science and Technology, Manchester, 1992.

[Lin76] J.G. Lin. Multi-objective problems: Pareto-optimal solutions by method of proper equality constraints. *IEEE Trans. Automatic Control*, AC-21(5):641–650, 1976.

[Lin77] J.G. Lin. Proper inequality constraints and maximization of index vectors. *J. Optim. Theory Applic.*, 21:505–521, 1977.

[Lin92] A. Linnemann. $\mathcal{L}_\infty$-induced optimal performance in sampled-data systems. *Syst. Control Lett.*, 18:265–275, 1992.

[Liu90] G. P. Liu. Frequency domain approach for critical systems. *Int. J. Control*, 52(6):1507–1519, 1990.

[Liu91] G. P. Liu. Input space and output performance. In *Proceedings of the American Control Conference*, pages 1197–1198, Boston, USA, 1991.

[Liu92a] G. P. Liu. Disturbance space and sup regulator in discrete time. *Syst. Control Lett.*, 18(1):33–38, 1992.

[Liu92b] G. P. Liu. Self-tuning sup regulators for discrete systems. In *Proceedings of the American Control Conference*, Chicago, USA, 1992.

[Liu93] G. P. Liu. Input space and output performance in the frequency domain for critical systems. *To appear in IEEE Trans. Automatic Control*, 1993.

[LLO87] R. Lozano-Leal and R. Ortega. Reformulation of the parameter identification problem for systems with bounded disturbances. *Automatica*, 23(2):247–251, 1987.

[LPF91] G.M.H. Leung, T.P. Perry, and B.A. Francis. Performance analysis of sampled-data control systems. *Automatica*, 27(4):699–704, 1991.

[Lue69] D. G. Luenberger. *Optimization by Vector Space Methods*. John Wiley & Sons, Inc., New York, 1969.

[LWT90] G. P. Liu, H. Wang, and J. X. Tian. The design of a hydraulic turbine control system — a critical system. *AMSE Periodicals: Modelling, Simulation and Control, B*, 33(3):23–34, 1990.

[LZ90a] G. P. Liu and V. Zakian. Sup regulators. In *The 29th IEEE Conference on Decision and Control*, pages 2145–2146, Hawaii, USA, 1990.

[LZ90b] G. P. Liu and V. Zakian. Sup regulators for critical systems. Control Systems Centre Report 731, University of Manchester Institute of Science and Technology, Manchester, U.K., 1990.

[Mac89] J.M. Maciejowski. *Multivariable Feedback Design*. Addison-Wesley, Wokingham, U.K., 1989.

[Mat87] The MathWorks, Inc., MA, USA. *PRO-MATLAB User's Guide*, 1987.

[McC83] G. P. McCormick. *Nonlinear Programming, Theory, Algorithm, and Applications*. John Wiley & Sons, New York, 1983.

[MK77] A.G.J. MacFarlane and B. Kouvaritakis. A design technique for linear multivariable feedback systems. *Int. J. Control*, 25:837–874, 1977.

[MPSV82] D.Q. Mayne, E. Polak, and A. Sangiovanni-Vincentelli. Computer-aided design via optimization: A review. *Automatica*, 18:147–154, 1982.

[Muh91] P.A. Muha. Expert system for SROOT. *IEE Proc. D*, 138(4):381–387, 1991.

[Mül77] P.C. Müller. Design of optimal state-observers and its application to maglev vehicle suspension control. In *Proc. 4th IFAC Symp. Multivariable Technological Systems*, pages 175–182, Fredericton, Canada, 1977.

[MZ89] M. Morari and E. Zafiriou. *Robust Process Control*. Prentice-Hall, Englewood Cliffs, N.J., 1989.

[NA86] K. S. Narendra and A. M. Annaswamy. Robust adaptive control in the presence of bounded disturbances. *IEEE Trans. Automatic Control*, AC-31(4):306–315, 1986.

[Ng89] W.Y. Ng. *Interactive Multi-Objective Programming as a Framework for Computer-Aided Control System Design*, volume 132 of *Lect. Notes Control & Inf. Sci.* Springer-Verlag, Berlin, 1989.

[NGK57] G.C. Newton Jr., L.A. Gould, and J.F. Kaiser. *Analytical Design of Linear Feedback Controls*. Wiley, New York, 1957.

[NM65] J.A. Nelder and R. Mead. A simplex method for function minimization. *Comp. J.*, 7(4):308–313, 1965.

[NT86] W.T. Nye and A.L. Tits. An application-oriented, optimization-based methodology for interactive design of engineering systems. *Int. J. Control*, 43(6):1693–1721, 1986.

[OLL87] R. Ortega and R. Lozano-Leal. A note on direct adaptive control of systems with bounded disturbances. *Automatica*, 23(2):253–254, 1987.

[Par06] V. Pareto. *Manuale di Economia Politica*. Societa Editrice Libraria, Milan, Italy, 1906.

[PB90] J.B. Pearson and B. Bamieh. On minimizing maximum errors. *IEEE Trans. Automatic Control*, AC-35(5):598–601, 1990.

[PHS+88] A.G. Parlos, A.F. Henry, F.C. Schweppe, L.A. Gould, and D.D. Lanning. Nonlinear multivariable control of nuclear power plants based on the unknown-but-bounded disturbance model. *IEEE Trans. Automatic Control*, AC-33:130–137, 1988.

[PM82] R.V. Patel and N. Munro. *Multivariable Theory and Design*. Pergamon Press, Oxford, England, 1982.

[Ros60] H.H. Rosenbrock. An automatic method for finding the greatest or least value of a function. *Comp. J.*, 3:175–184, 1960.

[Ros70] H. H. Rosenbrock. *State-Space and Multivariable Theory.* Nelson-Wiley, London, 1970.

[Ros74] H.H. Rosenbrock. *Computer-Aided Control Systems Design.* Academic Press, London, 1974.

[Rut92] N. Rutland. Illustration of a new principle of design: Vehicle speed control. *Int. J. Control*, 55(6):1319–1334, 1992.

[Sam83] C. Samson. Stability analysis of adaptively controlled systems subject to bounded disturbances. *Automatica*, 19:81–8, 1983.

[Sch73] F.C. Schweppe. *Uncertain Dynamical Systems.* Prentice-Hall, Englewood Cliffs, N.J., 1973.

[Sin87] P.K. Sinha. *Electromagnetic Suspension: Dynamics and Control.* Peter Peregrinus Ltd., London, 1987.

[SNF70] B. Sz.-Nagy and C. Foias. *Harmonic Analysis of Operators on Hilbert space.* North-Holland, Holland, 1970.

[Tai78] O. Taiwo. Improvement of turbo-alternator response by the method of inequalities. *Int. J. Control*, 27(2):305–311, 1978.

[Tai79] O. Taiwo. Design of multivariable controller for an advanced turbofan engine by Zakian's method of inequalities. *ASME J. Dyn. Syst. Meas. & Control*, 101:299–307, 1979.

[TSGJ79] D. Tabak, A.A. Schy, D.P. Giesy, and K.G. Johnson. Application of multiobjective optimization in aircraft control systems design. *Automatica*, 15:595–600, 1979.

[USWG82] P.B. Usoro, F.C. Schweppe, D.N. Wormley, and L.A. Gould. Ellipsoidal set-theoretic control synthesis. *ASME J. Dyn. Syst. Meas. & Control*, 104:331–336, 1982.

[Vid85a] M. Vidyasagar. *Control System Synthesis: A Factorization Approach.* MIT Press, Boston, MA., 1985.

[Vid85b] M. Vidyasagar. Filtering and robust regulation using a two-parameter controller. In *Proc. Sym. MTNS*, Stockholm, 1985.

[Vid86] M. Vidyasagar. Optimal rejection of persistent bounded disturbances. *IEEE Trans. Automatic Control*, AC-31:527–534, 1986.

[Vid91] M. Vidyasagar. Further results on the optimal rejection of persistent bounded disturbances. *IEEE Trans. Automatic Control*, AC-36:642–652, 1991.

[VJ84] M. S. Verma and E. Jonckheere. L^{∞}-compensation with mixed sensitivity as a broadband matching problem. *Syst. Control Lett.*, 4:125–130, 1984.

[WEPZ79] P. E. Wellstead, M. J. Edmunds, D. Prager, and P. Zanker. Self-tuning pole/zero assignment regulators. *Int. J. Control*, 30(1):1–26, 1979.

[Whi89] J.F. Whidborne. *Criteria - A Computer-Aided Control System Design Package: User Manual.* Control Systems Centre, UMIST, Manchester, 1989.

[Whi90] J.F. Whidborne. Time vs frequency domain characterisation of inputs. Control Systems Centre Report 732, University of Manchester Institute of Science and Technology, Manchester, U.K., 1990.

[Whi93] J.F. Whidborne. EMS control system design for a maglev vehicle - A critical system. *To appear in Automatica*, 1993.

[Wil70] J. L. Willems. *Stability Theory of Dynamical Systems.* Nelson-Wiley, London, 1970.

[WP86] Z. Z. Wang and J. B. Pearson. Regulation and optimal error reduction in linear multivariable systems. *Automatica*, 22:85–94, 1986.

[WPG92] J.F. Whidborne, I. Postlethwaite, and D.W. Gu. Robust controller design using H_∞ loop-shaping and the method of inequalities. Technical Report 92-33, Leicester University Engineering Department, Leicester, U.K., 1992.

[WPZ79] P. E. Wellstead, D. Prager, and P. Zanker. Pole assignment self-tuning regulator. *IEE Proc. D*, 126:781–787, 1979.

[YJB76] D. C. Youla, H. A. Jabr, and J. J. Bongiorno. Modern Wiener-Hopf design of optimal controllers – part 2: the multivarible case. *IEEE Trans. Automatic Control*,

AC-21:319–338, 1976.

[YP90] A. Yue and I. Postlethwaite. Improvement of helicopter handling qualities using H^∞-optimisation. *IEE Proc. D*, 137(3):115–129, 1990.

[Zad63] L.A. Zadeh. Optimality and non-scalar-valued performance criteria. *IEEE Trans. Automatic Control*, AC-8(1):59–60, 1963.

[Zak75] V. Zakian. Properties of I_{MN} and J_{MN} approximants and applications to numerical inversion of Laplace transforms and initial-value problems. *J. Math. Anal. Appl.*, 50(1):191–222, 1975.

[Zak78] V. Zakian. The performance and sensitivity of classical control systems. *Int. J. System Science*, 9(3):343–355, 1978.

[Zak79a] V. Zakian. Computation of abscissa of stability by repeated use of the Routh test. *IEEE Trans. Automatic Control*, AC-24(4):604–607, 1979.

[Zak79b] V. Zakian. New formulation for the method of inequalities. *Proc. IEE*, 126(6):579–584, 1979.

[Zak83] V. Zakian. A criterion of approximation for the method of inequalities. *Int. J. Control*, 37(5):1103–1111, 1983.

[Zak84] V. Zakian. A framework for design: Theory of majorants. Control Systems Centre Report 604, University of Manchester Institute of Science and Technology, Manchester, U.K., 1984.

[Zak86a] V. Zakian. A performance criterion. *Int. J. Control*, 43(3):921–931, 1986.

[Zak86b] V. Zakian. On performance criteria. *Int. J. Control*, 43(4):1089–1092, 1986.

[Zak87a] V. Zakian. Input spaces and output performance. *Int. J. Control*, 46(1):185–191, 1987.

[Zak87b] V. Zakian. Design formulation. *Int. J. Control*, 46(2):403–408, 1987.

[Zak89] V. Zakian. Critical systems and tolerable inputs. *Int. J. Control*, 49(4):1285–1289, 1980.

[Zak91] V. Zakian. Well matched systems. *IMA J. Control & Information*, 8:29–38, 1991.

[Zak92] V. Zakian. Principle of matching and the method of inequalities. Control Systems Centre Report 769, University of Manchester Institute of Science and Technology, Manchester, U.K., 1992.

[Zam81] G. Zames. Feedback and optimal sensitivity: model reference transformations, multiplicative seminorms, and approximate inverses. *IEEE Trans. Automatic Control*, AC-26(2):301–320, 1981.

[ZAN73] V. Zakian and U. Al-Naib. Design of dynamical and control systems by the method of inequalities. *Proc. IEE*, 120(11):1421–1427, 1973.

[ZD63] L.A. Zadeh and C.A. Desoer. *Linear Sytem Theory: The State Space Approach*. McGraw-Hill, New York, 1963.

[ZE78] V. Zakian and M.J. Edwards. Tabulation of constants for full grade I_{MN} approximants. *Math. of Comp.*, 32(142):519–531, 1978.

[ZF83] G. Zames and B. A. Francis. Feedback, minimax sensitivity, and optimal robustness. *IEEE Trans. Automatic Control*, AC-28(5):585–601, 1983.

Index